3/3/97

Other Titles in This Series

Stories about Maxima and Minima

V. M. Tikhomirov

Translated from the Russian by
Abe Shenitzer

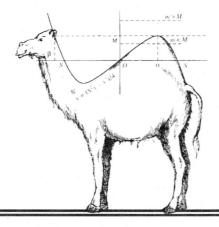

American Mathematical Society
Mathematical Association of America

В. М. ТИХОМИРОВ

РАССКАЗЫ
О МАКСИМУМАХ
И МИНИМУМАХ

«НАУКА», МОСКВА, 1986

Translated from the Russian by Abe Shenitzer

1991 *Mathematics Subject Classification.* Primary 00A07,
00A30, 00A35, 01–01, 46–01, 49–01, 49–03, 49J99

Library of Congress Cataloging-in-Publication Data

Tikhomirov, Vladimir M. (Vladimir Mikhaĭlovich), 1934–
    Stories about maxima and minima/V. M. Tikhomirov.
    p. cm.—(Mathematical world, ISSN 1055-9426; 1)
    ISBN 0-8218-0165-1
    1. Maxima and minima. 2. Calculus of variations. 3. Mathematical optimization.
QA306.T55 1990
511'.66—dc20

90-21246
CIP

To the Memory
of My Dear Friend,
V. M. Alekseev

Table of Contents

Introduction

In daily life it is constantly necessary to choose the best possible (optimal) solution. A tremendous number of such problems arise in economics and in technology. In such cases it is frequently useful to resort to mathematics.

In mathematics, the study of maximum and minimum problems began a very long time ago, in fact, twenty-five centuries ago. For a long time there were no uniform ways of tackling problems for finding extrema. The first general methods of investigation and solution of extremal problems were created about 300 years ago, at the time of the formation of mathematical analysis.

Then it became clear that certain special optimization problems play a crucial role in the natural sciences. Specifically, it was found that many laws of nature can be derived from so-called "variational principles." According to these principles, given any collection of admissible motions, what distinguishes the actual motion of a mechanical system, or of light, electricity, a fluid, a gas, and so on, is that it maximizes or minimizes certain quantities. Some concrete extremal problems, whose content derives from the natural sciences (the brachistochrone problem, Newton's problem, and others), were posed at the end of the seventeenth century. The need to solve these, as well as many other problems of geometry, mechanics, and physics, led to the creation of a new branch of mathematical analysis that came to be known as the calculus of variations.

The intensive development of the calculus of variations continued for about two centuries. Many of the finest scientists of the eighteenth and nineteenth centuries took part in this process, and, by the beginning of this century, it seemed as if they had exhausted the topic.

But it turned out that this was not the case. The needs of practical life, especially in economics and technology, gave rise to new problems that could not be solved by the old methods. One had to advance. It was necessary to create a new field of mathematical analysis, known as "convex analysis," involving the study of convex functions and convex extremal problems.

The needs of technology, and in particular the exploration of space, gave rise to yet another series of problems that were likewise unsolvable by the methods of the calculus of variations. Thus, another new theory, known as optimal control theory, was created. The fundamental method of optimal control theory was worked out in the 1950s and 1960s by Soviet mathematicians, namely L. S. Pontryagin and his colleagues. This provided a new and powerful impulse for further investigations in the theory of extremal problems.

This book aims to acquaint the reader with this whole circle of ideas. However, this is not the author's only purpose. Throughout the history of mathematics, maximum and minimum problems have played an important role in its evolution. During this time many beautiful, important, brilliant, and interesting problems in geometry, algebra, physics, and so on, have appeared. The greatest scientists of the past—Euclid, Archimedes, Heron, Tartaglia, Johann and Jakob Bernoulli, Newton, and many others—took part in the solution of these concrete problems. The solutions stimulated the development of the theory and, as a result, techniques were elaborated that made possible the solution of a tremendous variety of problems by a single method.

The author would like the reader to understand how and why a mathematical theory is born. In Part One, the reader will get to know many concrete problems, and in the course of the discussion of their solutions he will come in contact with the creative work of some of the best mathematicians of the past. This is not only of historical interest. For the most part, the ideas and methods created by eminent mathematicians in connection with the solution of problems do not die and are certain to be reborn, given enough time. That is why to fathom the conceptions of great men is always an enriching experience.

The need to solve a large number of varied problems establishes the preconditions for the creation of a general theory. In Part Two I will introduce a method for solving maximum and minimum problems that originated with Lagrange. The basic conception of this method has endured for over two centuries. Its content has varied constantly, but its key thought has remained unchanged. It is not a simple matter to understand the reasons for this universality of Lagrange's idea. On the other hand, it is not at all difficult to learn to use Lagrange's principle for the solution of problems. At the end of Part Two all problems discussed in Part One, problems marked by the dissimilarity of their solutions, are investigated and solved by means of a single general method, in a standard way, using one and the same scheme.

The author has tried to show how the analysis of diverse facts gives rise to a general idea, how this idea is transformed, how it is enriched by new content, and how it remains the same under all changes.

With the exception of the concluding part of the fourteenth story, this book is primarily aimed at high school students. But I would very much like its readers to include college students interested in mathematics and, of course, teachers. The last story is addressed above all to them. It impinges on the question of how and why to teach. I think that the content of the book supplies material that is ideally suited for a discussion of this topic, a topic that is bound to concern us for many years to come. Thus, I hope that this book will also be read by my colleagues who study mathematics and teach it to their students.

I wish to thank all those who read the manuscript and commented on it. This refers, above all, to Andrei Nikolaevič Kolmogorov, Nikolai Borisovič Vasil′ev, Ivan Penkov, and Georgiĭ Georgevič Magaril-Il′yaev.

I am grateful to Prof. E. Barbeau for a number of valuable remarks that have been included in the English translation of my book. I also wish to express my deep appreciation to Prof. A. Shenitzer for his work as translator.

V. M. Tikhomirov

PART ONE

Ancient Maximum and Minimum Problems

Mathematics...possesses not only truth, but supreme beauty...such as only the greatest art can show.

B. Russell

The most fascinating pursuit is to follow the thoughts of a great man.

A. S. Pushkin

1

Why Do We Solve
Maximum and Minimum Problems?

Nothing takes place in the world whose meaning is
not that of some maximum or minimum.

L. Euler

Most practical questions can be reduced to problems
of largest and smallest magnitudes...and it is only by
solving these problems that we can satisfy the requirements of practice which always seeks the best, the
most convenient.

P. L. Čebyšev

...one wants to reach the very essence.

B. L. Pasternak

We learn about maxima and minima in school. One ancient problem that
you may have solved in your geometry lessons is the following:

*A and B are two given points on the same side of a line l. Find a point
D on l such that the sum of the distances from A to D and from D to B
is a minimum* (Figure 1.1 on page 4).

Here it is necessary to find a least value, that is, a *minimum*. In many
problems it is necessary to find a *maximum*, the largest value of

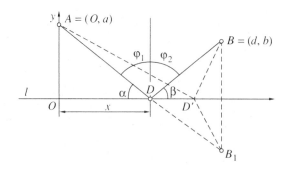

FIGURE 1.1

something. Both notions—maximum and minimum—are subsumed under the Latin term *extremum*. Problems that involve finding maxima and minima are called *extremal problems*. (The term *optimization problems* has almost the same meaning.) The methods of solution and investigation for the various extremal problems constitute distinct chapters of mathematical analysis. Together, these methods make up the part of analysis called the *theory of extremal problems*.

Our aim in this book is to consider two questions: Why do we solve maximum and minimum problems? What are the components of the theory of extremal problems?

Earlier we posed a geometric problem. This problem can be found in almost all geometry textbooks. When and why did this problem first appear?

The presumed author is the famous ancient mathematician Heron of Alexandria. (In this text we will call the problem *Heron's problem*.) We all know about Heron through the formula for the area of a triangle that bears his name. The book containing this problem is titled, *On mirrors*. Scholars disagree as to when this book was written, but most believe that it was written in the first century A.D. Although Heron's book has disappeared, we know about it from later commentaries.

I assume that the reader knows about Heron's problem and has solved it. I stated it because it will be very useful for illustrating various points.

Let's recall the solution of Heron's problem.

Let B_1 be the point symmetric to B with respect to the line l. Join A to B_1. *The required point D is the point of intersection of AB_1 and l* (see Figure 1.1). Indeed, if D' is a point other than D, then

$$(1) \qquad |AD'| + |D'B| = |AD'| + |D'B_1| > |AB_1| = |AD| + |DB|.$$

Here and in the sequel $[AB]$ denotes the segment joining the points A and B, $|AB|$ denotes the length of $[AB]$ and $AB\|CD$ indicates that the lines AB and CD are parallel.

In establishing (1) we made use of symmetry properties that imply the equalities $|DB| = |DB_1|$, $|D'B| = |D'B_1|$, and the triangle inequality $|AD'| + |D'B_1| > |AB_1|$. This completes the solution of the problem.

We note that the required point D has the property that the angle α is equal to the angle β. (See Figure 1.1.) Also, the angle φ_1 is equal to the angle φ_2, or, as is usually said, *the angle of incidence is equal to the angle of reflection*.

Using the idea in the argument just presented, try to solve the following problems.

PROBLEM 1. *Let C be a given point in the interior of a given angle. Find points A and B on the sides of the angle such that the perimeter of the triangle ABC is a minimum.*

PROBLEM 2. *Given an angle and two points C and D in its interior, find points A and B on the sides of the angle such that $|CA| + |AB| + |BD|$ is a minimum.*

Let's return to Heron's problem. In his book Heron investigates the laws of reflection of light and applies his conclusions to problems related to properties of mirrors. In particular, he proves that a parabolic mirror brings to a focus the pencil of rays parallel to the mirror's axis.

In Heron's time scholars tried to comprehend the laws of nature by speculation and logical arguments, without recourse to experiment. Later in this book we will have occasion to talk of the rise of modern experimental science. The first great experimenter in the history of science was Galileo Galilei, who lived in the seventeenth century. In contrast to Galileo, Heron tried to base his explanations of the laws of reflection on logical foundations. He seems to have assumed that nature pursues the shortest path. Damianus (sixth century A.D.), a commentator on Heron, says that

> Heron...showed that lines inclined at equal angles are the
> smallest of all intermediate ones inclined on the same side
> of a single line. Proving this, he says that if nature does not
> want a ray of light to meander to no purpose, then it breaks
> it at equal angles.

Historians of science see in this the first hint of the thought that nature is guided by *extremal principles*. Heron's idea was developed further by Fermat (we will have more to say about this in our third story). Fermat deduced the law of refraction of light (established earlier experimentally by Snel) from the assumption that what characterizes the trajectory of a light ray moving from one point to another in a nonhomogeneous medium is that it is traversed in a minimum of time. From that point on, the idea of the extremal character of natural phenomena became the guiding light of science. This is confirmed by the words of Euler that we chose as an epigraph for this story.

I will postpone the discussion of the remarkable character of this phenomenon—after all, we cannot think of nature as having a purpose. Nevertheless, what distinguishes the trajectories of light and radio waves, the motions of pendulums and planets, the flows of liquids and gases, as well as many other motions, is that they are all solutions of problems of maxima and minima. This fact provides a fruitful means of creating a mathematical description of nature.

This, then, is the main reason that impels us to solve problems of maxima and minima and to develop the theory of extremal problems. It gave rise in the eighteenth century to a special part of this theory called *the calculus of variations.*

Another reason for studying these problems is found within ourselves. Human beings constantly strive to better themselves, which is why they always want to choose the best of existing possibilities. In this endeavor, mathematics can sometimes be of help.

Let's discuss this again using Heron's problem as a relevant example. Some textbooks state it as a practical problem. The line l is turned into a rectilinear section of railroad track, points A and B become towns, point D is called a railroad platform, and the question is: *where should one build the platform so that the combined length of the rectilinear highways linking it to the towns is minimal?*

What follows are some additional geometric problems of possible practical value. Try to think them through yourself.

PROBLEM 3. *Let A, B, and C be three towns. Find D such that the combined length of the rectilinear highways linking it to A, B, and C is minimal.*

PROBLEM 4. *Solve Problem 3 for four towns.*

PROBLEM 5. *Will the answer to Problem 4 change if we ask for the minimal length of highway linking the four towns without specifying that the highway links must come together in one point?*

It is clear that such problems are just models of actual situations. In reality, all is far more complicated: sections of railroad tracks are not rectilinear, highways are not built to follow strictly straight lines, and a "sum of distances" alone is seldom an "optimality criterion." But there is no doubt that in building railroad tracks, highways or other roads, gas and oil pipelines, and in many other situations, the usual question is how to accomplish the task most expediently—say, at least cost.

Such problems arise constantly in economic activities. Invariably, an objective must be attained in the cheapest, fastest, shortest, or most economical manner.

Let's look at an optimization problem in an economic context. Suppose we have supply centers of a certain product, stores, and a truck depot. How should the truck depot dispatcher organize the supplying of the necessary product to the stores for maximum economy? (Problems of this type are called transportation problems. We will formulate them more precisely later.) To solve such problems, it is necessary to turn to mathematics.

The methods for solving maximum and minimum problems developed through the middle of the present century proved inadequate for the solution of problems similar to our transportation example. One of the things that came to light is that in many economic problems the notion of *convexity* plays a key role. Since one often encounters convex as well as linear functions and sets in this area, the need arose for a comprehensive theory of convex sets and functions now known as *convex analysis*. This development gave rise to new directions in the theory of extremal problems called *linear and convex programming*. They were initiated in the 1930s by the Soviet mathematician L. V. Kantorovič.

Most optimization problems deal with technological processes, tools, and systems. Here is a relevant example. Consider a cart moving rectilinearly and without friction on horizontal rails. The cart is controlled by an external force that can be varied within prescribed bounds. The cart is to be stopped at a definite location in the shortest possible time. This problem exemplifies *the simplest problem of rapid response under automatic control*. It is an instance of the multitude of problems that have arisen in the chemical industry, in space travel, and in other technological areas that could not be handled by the methods of the calculus of variations. Thus, it became necessary to create a new field to supplement the calculus of variations. This new field was called *optimal control*.

All this points to yet another reason for the solution of optimization problems and the development of the theory of extremal problems. To quote Čebyšev, it is the wish "to satisfy the requirements of practice." But these reasons do not explain the whole mystery.

The next story deals with the oldest maximum and minimum problem, namely the classical isoperimetric problem. Some twenty-five centuries ago, in ancient Greece, it was discovered that of all closed curves of a given length, the circle has the remarkable property of enclosing the largest area. In school you probably encountered problems describing analogous properties of polygons. Let's recall two such exercises.

PROBLEM 6. *Find a triangle of given perimeter that has maximal area.*

PROBLEM 7. *Show that of all rectangles of given perimeter, the square has the largest area.*

A problem equivalent to Problem 7 is dealt with already in Euclid's *Elements*. Also, Fermat used the solution of this very problem to illustrate his method of finding maxima and minima, a method known as Fermat's theorem.

Why were such problems posed and solved? What is the secret of their attraction? Why do the authors of most geometry books like to deal with problems of maxima and minima?

These questions are not easily answered. The fact remains, however, that throughout the history of mathematics, extremal problems have elicited interest and a desire to solve them. It is conceivable that what is behind this is our natural quest for perfection, some secret tendency to comprehend "the very essence." Perhaps it is that most—if not all—extremal problems contain an element of grace, of attractiveness, of the beauty of which Russell speaks, and this impels us to solve maximum and minimum problems.

I have said enough for you to appreciate the importance and interest of the subject I have chosen.

Perhaps it is relevant to indicate the temporal bounds of our stories. The earliest maximum and minimum problems were posed in the distant past. In fact, the classical isoperimetric problem covered in the next story was investigated in the fifth century B.C. And in the fourteenth story we will deal with problems arising in our own time.

For a long time, each extremal problem was solved individually. In the seventeenth century, there was a clear awareness of the need to create some general methods. Such methods were developed by Fermat, Newton, Leibniz, and others; first for one, then for finitely many, and, ultimately, for infinitely many variables. These methods led to the formulation of the basic divisions of the theory of extremal problems: *mathematical programming* (that is, the theory of finite-dimensional optimization problems), *convex* (including linear) *programming* (where one studies convex optimization problems), *the calculus of variations, and the theory of optimal control*.

This book is divided into two parts. The first part consists of ancient problems, posed and solved, as a rule, before the invention of the first general methods. In the second part we will discuss some of the methods of the theory of extremal problems.

In the first part we will discuss problems connected with the names of the greatest mathematicians of various epochs such as Euclid, Archimedes, Fermat, Kepler, Huygens, Johann Bernoulli, Newton, and Leibniz. I have not denied myself the pleasure "of following the thoughts" of these great men.

And in the second part.... Of that it is as yet too early to talk.

2

The Oldest Problem—Dido's Problem

They bought as much land—and called it Birsa—as
could be encircled with a bull's hide.

The Aeneid of Vergil

The most beautiful solid is the sphere, and the most
beautiful plane figure—the circle.

Pythagoras

We took as an epigraph for this story two lines from the *Aeneid* of Publius
Vergilius Maro, one of the greatest poets of ancient Rome. Like all immortal
creations, the *Aeneid* tells the story of human passions, of good and evil,
of fate and suffering, of guile and love, of life and death. The quoted lines
refer to an event that tradition placed in the ninth century B.C. We recall the
legend reproduced in the *Aeneid*.

Fleeing from persecution by her brother, the Phoenician princess Dido set
off westward along the Mediterranean shore in search of a haven. A certain
spot on the coast of what is now the bay of Tunis caught her fancy. Dido
negotiated the sale of land with the local leader, Yarb. She asked for very
little—as much as could be "encircled with a bull's hide." Dido managed to
persuade Yarb, and a deal was struck. Dido then cut a bull's hide into narrow
strips, tied them together, and enclosed a large tract of land. On this land
she built a fortress and, near it, the city of Carthage. There she was fated to
experience unrequited love and a martyr's death.

This incident suggests the question: How much land can be enclosed by a
bull's hide?

Why begin with this problem? After all, its solution is rather difficult. It would seem reasonable to begin with simpler matters. Still, I choose another road. In this part I will move not from simple to complex matters, but from the distant past to our own days. That is why I want "to begin at the beginning." Is it not remarkable that such difficult and profound problems were posed and solved in those mythical times? Our predecessors knew so much less than we do, but they persevered and attained their objective!

How much land, then, can one enclose by a bull's hide? To answer this question, we must pose it in a mathematically correct manner. A modern mathematician would say:

Among all closed plane curves of a given length, find the one that encloses the largest area.

This question is known as *Dido's problem*, or the *classical isoperimetric problem*. (Isoperimetric figures are figures that have the same perimeter.)

So far we have managed with words alone. A person with a sufficiently high level of mathematical culture is completely satisfied with this kind of formulation, for he knows what is meant by "curve," "length," and "area." It took more than 2000 years to assign precise meanings to these words. To properly clarify these terms would require another book, so we will approach our problem in the "naive" manner of the ancients (and in the manner dictated by practical considerations to Princess Dido herself). But we will try to do without a bull's hide.

We unwind some thread from a spool, cut it, tie the ends together, and put the tied thread on a sheet of paper. The result is *a plane closed curve*. If we now cut out the piece of paper along the contour of the thread, then we obtain a representation *of the area enclosed by this curve*. This area can be measured. If our sheet of paper is a sheet of millimeter paper, then the measurement can be quite accurate. Now the question posed by the problem is clear: we are to explain *how to place our thread to enclose a maximum area*.

I will soon show that the curve that solves the classical isoperimetric problem is a circle. In describing Dido's actions, Vergil used the word "circumdare" (to encircle) containing the root *circus* (circle). This suggests that Dido solved the classical isoperimetric problem correctly.

Many historians are of the opinion that this was the first extremal problem discussed in the scientific literature. In addition to noting the isoperimetric property of the circle (that is, the property of the circle to enclose the largest area among all isoperimetric figures), ancient geometers also noted the *isoepiphanic property of the sphere* (that is, the property of the sphere to enclose the largest volume among all figures with the same surface area). This property of maximal capacity was the basis of the notion that the circle and sphere are the embodiments of geometric perfection (recall the words of Pythagoras that serve as an epigraph for this story).

Another confirmation of the same thought is found in the words of Nicolaus Copernicus:

> In the first place we must observe that the universe is spherical. This is either because that figure is the most perfect, as not being articulated, but whole and complete in itself; or because it is the most capacious and therefore best suited for that which is to contain and preserve all things...

It is now impossible to tell when the thought of the maximal capacity of the circle and the sphere was first advanced. At any rate, Aristotle (4th century B.C.)—one of the greatest thinkers in human history—treats these facts as given. And who (other than Dido) did, in fact, solve the isoperimetric problem? The literature devoted to the isoperimetric property of the circle and the isoepiphanic property of the sphere is vast. One of the immense number of these works is by the German geometer W. Blaschke [2], which includes historical references. Should you be tempted to follow "the history of the isoperimetric problem "from its beginning in hoary antiquity with the legend of the Carthaginian princess Dido to *Herr Geheimrat* Hermann Amandus Schwartz from Berlin" [1] you could turn to Blaschke's paper [2].

One of the presumed solvers of the isoperimetric and isoepiphanic problems mentioned by the ancient authors is Archimedes. H. A. Schwartz is thought to have given the first *rigorous* proofs of the maximum property of the circle and the sphere.

But in fact, Schwartz—and before him Weierstrass, and after him Blaschke himself, and numerous other mathematicians in the nineteenth and twentieth centuries—should be given credit (in connection with the isoperimetric problem) merely for shaping the ideas of their distant predecessors so as to meet the requirements of rigor of their time. The basic ways of solving the isoperimetric problem were already outlined with absolute correctness in ancient times. We will now describe one such way, due to Zenodorus, a mathematician who is thought to have lived sometime between the third century B.C. and the first century A.D.

Zenodorus proves completely rigorously—by the standards of his time—the following assertion.

If there exists a plane n-gon having largest area among all n-gons of given perimeter, then it must have equal sides and equal angles.

In the interest of brevity, we will call a plane *n*-gon of largest area, among all *n*-gons isoperimetric with it, a *maximal n-gon*. Using this term we can state Zenodorus' theorem more briefly.

A maximal n-gon (if one exists) must be regular.

Zenodorus' theorem follows from two lemmas.

LEMMA 1. *A maximal n-gon must have equal sides.*

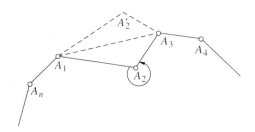

FIGURE 2.1

LEMMA 2. *A maximal n-gon must have equal angles.*

When presenting the works of our distant mathematical predecessors, I will not, as a rule, reproduce them literally, preserve the notation and style of the authors, or strive to give the authors' own proofs. Instead, I will reproduce their basic direction of thought and general spirit of argument, while changing and modernizing formulations and proofs. In particular, I will present modified proofs of Lemmas 1 and 2. I will use the solution of Heron's problem twice.

Before presenting the proofs, it is necessary to make an observation not mentioned by Zenodorus. As we are about to show, a *nonconvex polygon cannot be maximal*. Indeed, suppose that the angle $A_1A_2A_3$, say, is larger than $180°$. (See Figure 2.1) Let A_2' be the image of the vertex A_2 under reflection in the line A_1A_3. The polygon $A_1A_2'A_3 \ldots A_n$ has greater area than the polygon $A_1A_2A_3 \ldots A_n$ and is isoperimetric with it. Now we are ready to give:

PROOF OF LEMMA 1. Let $A_1A_2 \ldots A_n$ be a maximal n-gon. As noted, it is a convex figure. We suppose that not all of its sides are equal and deduce a contradiction.

Let A_1A_2 and A_2A_3 be two adjacent unequal sides. Let l be the line through A_2 parallel to A_1A_3. (See Figure 2.2.) Now consider Heron's problem for the line l and the points A_1 and A_3. Recall that this is the problem of finding a point D on l that minimizes the sum of the distances $|A_1D| + |A_3D|$. As was proved in the previous section, the angles α and β at D must be equal. But α is equal to the angle DA_1A_3, and β is equal to the angle DA_3A_1 (by the property of opposite alternate angles between parallels). This means that A_1DA_3 is an isosceles triangle, and therefore D is different from A_2. Furthermore,

(a) the area of $\triangle A_1DA_3$ is equal to the area of $\triangle A_1A_2A_3$, since they have equal altitudes and bases; and

(b) the sum of the sides A_1D and DA_3 is less than the sum of the sides A_1A_2 and A_2A_3, since D ($\neq A_2$) is the solution of Heron's problem.

We now construct the isosceles triangle $A_1A_2'A_3$ such that $|A_1A_2'| + |A_2'A_3|$

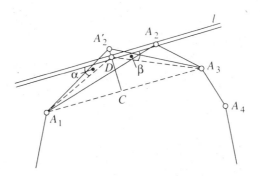

FIGURE 2.2

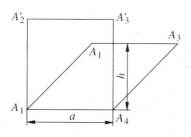

FIGURE 2.3

$= |A_1A_2| + |A_2A_3|$. Its area is, of course, larger than the area of $\triangle A_1A_2A_3$, since the altitude $A_2'C$ is larger than the altitude DC (by virtue of the fact that $|A_1A_2'|$ is longer than $|A_1D|$). But this means that the area of the polygon $A_1A_2' \cdots A_n$ is greater than the area of the polygon $A_1A_2 \cdots A_n$ isoperimetric with it, a conclusion that contradicts the maximality of the latter polygon. This completes the proof of Lemma 1.

COROLLARY. *Lemma 1 implies that a maximal triangle is equilateral and a maximal quadrilateral is a rhombus.*

This corollary and Figure 2.3 justify the conclusion that a maximal quadrilateral is, in fact, a square.

PROOF OF LEMMA 2. Again, let $A_1A_2 \cdots A_n$ be a maximal polygon. We know by now that all its sides are equal (Lemma 1) and bear in mind that it must be convex. We will suppose that not all of its angles are equal and will deduce a contradiction. If the angles are not all equal, then there must be two unequal adjacent angles, α and β, say. We will show that this implies the existence of two unequal nonadjacent angles.

Consider the successive angles α, β, γ, δ, ε, ... (there are no fewer than five) of the polygon. If $\gamma \neq \alpha$ or $\delta \neq \beta$, then the proof is complete, since α and γ (or β and δ) are nonadjacent. If $\alpha = \gamma$, $\beta = \delta$, and $\alpha \neq \beta$, then our sequence of angles is α, β, α, β, ε ... , and the proof is complete, since

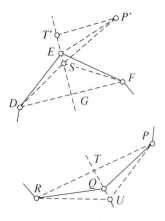

FIGURE 2.4

the first and fourth angles are nonadjacent.

We see that our assumption justifies the conclusion that there are two triangles DEF and PQR with disjoint interiors (Figure 2.4), each of which is formed by successive vertices of our n-gon and such that angle E is smaller than angle Q. Since $|DE| = |EF| = |PQ| = |QR|$, the inequality of the angles E and F implies that $|DF| < |PR|$. From E and Q we drop perpendiculars EG to DF and QT to PR. Next, we extend the segment EG and apply to the extension the triangle $ET'P'$ congruent to the triangle QTP (T goes over into T', P into P' and Q into E). Now we consider Heron's problem for the line $T'G$ and the points P' and F. Let S be the solution of Heron's problem, that is S is a point on $T'G$ such that the sum of the distances from P' to S and from S to F is minimal. Since the angle $P'ET'$ (equal to half the angle Q) is larger than the angle FEG (equal to half the angle E), the point S does not coincide with the point E (the angles $P'ST'$ and FSG are equal) and, furthermore, S lies on the segment EG. Now we lay off on the line QT the segment TU of the same length as the segment $T'S$ and consider the triangles DSF and PUR. The sum of the lateral sides of these triangles is smaller than the sum of the lateral sides of the original triangles DEF and PQR. In fact,

$$|DS| + |SF| + |PU| + |UR| = 2(|SF| + |SP'|) < 2(|FE| + |EP'|)$$
$$= |DE| + |EF| + |PQ| + |QR|.$$

We have used the fact that our triangles are isosceles and that S is the solution of Heron's problem. On the other hand, the area of $\triangle P'ES$ is larger than the area of $\triangle ESF$, since their respective altitudes are $|P'T'| = \frac{1}{2}|PR|$ and $|FG| = \frac{1}{2}|DF|$ and we have shown that $|DF| < |PR|$. It follows that the sum of the areas of the triangles DSF and PUR is greater than the sum of the areas of the original triangles DEF and PQR. In fact, denoting the area of a triangle UVW by $S_{\triangle\, UVW}$, we have

(2).
$$S_{\triangle\, DSF} + S_{\triangle\, PUR}$$
$$= S_{\triangle\, DEF} - 2_{\triangle\, ESF} + S_{\triangle\, PQR} + 2S_{\triangle\, P'ES} > S_{\triangle\, DEF} + S_{\triangle\, PQR}.$$

This means that the polygon $DSF\ldots PUR\ldots$ has a smaller perimeter and a larger area than our original polygon $DEF\ldots PQR\ldots$. Now we can treat either triangle (DSF or PUR) as we treated $\triangle\, A_1DA_3$ in proving Lemma 1, that is, we can raise it to obtain a polygon isoperimetric with the polygon $DEF\ldots PQR\ldots$. Since the area of the new polygon is greater than the area of the polygon $DSF\ldots PUR\ldots$, it is certainly greater than the area of the polygon $DEF\ldots PQR\ldots$. This contradicts the maximality of the polygon $DEF\ldots PQR\ldots$ and completes the proof of Lemma 2 and, thereby, also of the theorem of Zenodorus.

It remains to deduce from this theorem a proof of the classical isoperimetric theorem.

Lemma on the existence of a maximal n-gon. We have shown that if a maximal n-gon exists then it must be regular. But does a maximal n-gon exist? If it doesn't, the solution of Dido's problem turns to dust and ashes. After all, not all functions attain a maximum. For example, the function $f(x) = -(1 + x^2)^{-1}$ doesn't (this example is analyzed in greater detail in the eleventh story).

The ancient authors did not concern themselves with questions of existence of solutions. It was only some 100 years ago that mathematicians began to appreciate the significance of existence questions and to develop methods of proof of existence theorems. Later we will have many occasions for dealing with these questions. Here we will state without proof the following assertion (whose truth seems to have been obvious to Zenodorus).

LEMMA 3. *There exists a maximal n-gon.*

This and Lemmas 1 and 2 imply:

THEOREM 1. *A maximal n-gon is regular.*

Now there is little left to prove.

COMPLETION OF THE PROOF. Let P denote the perimeter of a regular n-gon and S its area. We know from geometry that $P = 2nR\sin(\pi/n)$, where R is the radius of the circumscribed circle, and that $S = rP/2$, where r is the radius of the inscribed circle. We have $r = R\cos(\pi/n)$. All these yield the following formula linking S and P:

$$P^2 - 4n\tan(\pi/n)S = 0.$$

Theorem 1 implies that if P is the perimeter of an arbitrary n-gon and S is its area, then

(3) $$P^2 - 4n\tan(\pi/n)S \geq 0.$$

The inequality $\tan \alpha \geq 0$ (valid for $0 < \alpha \leq \pi/2$) and (3) imply the inequality

$$(4) \qquad\qquad\qquad P^2 - 4\pi S \geq 0,$$

which holds for an arbitrary n-gon and all n. We note that for an arbitrary circle we have the obvious equality

$$(5) \qquad\qquad\qquad P^2 - 4\pi S = 0,$$

where P is the circumference of the circle and S is its area.

Now we will state a lemma linking together all the concepts involved in the formulation of the classical isoperimetric problem and the notion of an n-gon. Its meaning is that it is possible to approximate the length of a curve and the area it enclosed by means of the length and area of an n-gon, and to do so with arbitrary precision.

LEMMA 4. *For every closed plane curve of length P^* that enclosed an area S^* and for every $\varepsilon > 0$, there is an n-gon of perimeter P and area S such that*

$$(6) \qquad\qquad\qquad |P - P^*| \leq \varepsilon, \ |S - S^*| \leq \varepsilon.$$

Lemma 4 and the relation (4) imply that for every ε there is a polygon with perimeter P and area S such that

$$4\pi S^* \leq 4\pi S + 4\pi\varepsilon \leq P^2 + 4\pi\varepsilon \leq (P^* + \varepsilon)^2 + 4\pi\varepsilon = P^{*2} + \varepsilon(2P^* + 4\pi + \varepsilon).$$

Since ε is arbitrary, we arrive at the final inequality

$$(7) \qquad\qquad\qquad 4\pi S^* \leq P^{*2}.$$

According to (5), this inequality becomes an equality for a circle.

We sum all this up in the following theorem.

THEOREM 2. *The area enclosed by an arbitrary closed curve of given length does not exceed the area enclosed by a circle of the same length.*

This completes the solution of the isoperimetric problem.

COMMENTS. 1. We obtained a complete solution of our problem by combining the two geometric lemmas of Zenodorus and the two modern, essentially technical, Lemmas 3 and 4. All information necessary for a proof of Lemma 3 is to be found in the works of Weierstrass. The notions of the length of a curve and of the area enclosed by a curve were made precise by Jordan, who thereby provided the basics for a proof of Lemma 4.

2. Detailed proofs of Lemmas 3 and 4 can be found in Blaschke [2].

Before ending this story we will digress one last time.

Steiner's proof. Having presented a proof based on the ideas of the ancients, it is difficult to resist presenting an outline of yet another proof, whose

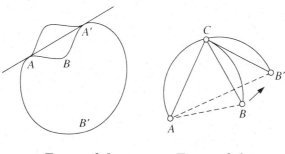

FIGURE 2.5   FIGURE 2.6

key thought is due to Jakob Steiner, a mathematician who enriched geometry with many remarkable ideas. A tacit assumption of Steiner's proof is the existence of the curve that solves the isoperimetric problem. (We already know that this is a justified assumption.) It remains to show that this extremal curve is a circle.

ASSERTION 1. *The extremal curve is convex.*

What is a convex curve? It is a curve whose interior (that is, the region bounded by the curve) includes the segment joining any two of its points.

In this connection, we wish to note that convexity plays a key role in maximum and minimum problems. We will have things to say about it in the sequel. Many remarkable books dealing with convexity are intended for high school students. One such book is by Lyusternik [7R] and another by Yaglom and Boltyanskii [13].

Let's turn to the proof of Steiner's theorem and prove Assertion 1.

If the curve is not convex then it must contain two points A and A' such that both arcs ABA' and $AB'A'$ joining A and A' lie on the same side of the line AA'. (See Figure 2.5.) By replacing one of these arcs with its image under reflection in AA', we obtain a new curve of the same length that encloses a larger area.

ASSERTION 2. *If points A and B halve the length of the extremal curve, then the chord $[AB]$ halves the area it encloses.*

In fact, if the chord $[AB]$ divided the area into unequal parts, then the figure consisting of the larger part and its image under reflection in the diameter AB would add up to a figure with the same length and a larger area.

ASSERTION 3. *Suppose that points A and B halve the extremal curve. If C is any point on the curve, then the angle ACD is a right angle.*

This is the heart of the matter. The method we will employ to prove this assertion is known as *Steiner's hinged-quadrilateral method*.

Suppose there is a point C such that the angle ACB is not a right angle. The area bounded by the arc ACB and the diameter AB splits into three parts, namely the triangle ABC and the segments adjacent to the sides AC and CB. Now imagine that there is a hinge at C linking together the two segments. "Spread" the segments so that the angle ACB' is a right angle. (See Figure 2.6 on page 17.) The area bounded by the arc ACB' will have increased, because, of all triangles with given lateral sides the right triangle has maximal area ($S_{\triangle\,ABC} = \frac{1}{2}|AC||BC|\sin C \leq \frac{1}{2}|AC||BC|$ and equality is attained if the angle is $90°$). The figure obtained by reflecting the curve ACB' in the chord AB' has the same perimeter but a larger area than the original figure. This proves our assertion.

We see that *the extremal figure consists of all points C from which a chord that halves the length of the extremal curve is seen at a right angle—that is, the curve in question is a circle.*

An enthusiast will exclaim: "Astounding!" A skeptic will nag: "This hasn't been shown, that must be justified.... Try to prove existence.... How do we know that when we spread the hinge far enough, parts of the segments at C won't overlap?" We'll ignore his grumbling. Granted, the proof is amazing, but it must be justified!

Of the many books dealing with the isoperimetric problem, I recommend Courant and Robbins [3], Kryžanovskiĭ [6R], and Rademacher and Toeplitz [12] for further reading.

3

Maxima and Minima in Nature (Optics)

According to Leibniz our world is the best possible. That is why its laws can be described by extremal principles.

C. L. Siegel

Carl Siegel, an eminent twentieth-century mathematician, obtained fundamental results in many areas of mathematics and mechanics. His remark that is the epigraph for this story is a joke, of course, but it contains a kernel of truth. When discussing Heron's problem, we had cause to remark that nature "employs" extremal principles. For example, we said that a reflection from a flat surface "chooses" a trajectory of least length.

Heron's words quoted in the first story contain the germ of a fundamental idea established between the seventeenth and nineteenth centuries. During this time it became clear that nature "operates" optimally in optics, in mechanics, in thermodynamics—in fact, everywhere.

The extremal principle associated with natural phenomena was clearly formulated for the first time in optics in an attempt to comprehend the law of refraction of light. The book of Tarasov and Tarasova [9R] deals with various optical problems and, in particular, with the history of the law of refraction.

The refraction of light is readily apparent in nature. For example, a pole lowered into a calm, transparent lake looks bent as a result of this phenomenon.

Ancient philosophers tried to discover the law of refraction. In particular, in the second century B. C. Ptolemy tried to obtain this law experimentally but failed to do so.

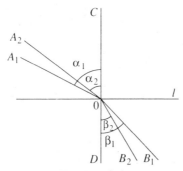

FIGURE 3.1

The law was first found by the Dutch scientist Snel. Snel's name is not as well known today as the names of his great contemporaries Descartes, Huygens, and Fermat. Snel's fame only results from his experimental discovery of the law of refraction of light, a discovery that remained unpublished in his lifetime. In his time Snel was very famous. Kepler regarded him as "the glory of the geometers [mathematicians] of our age."

Snel's law of refraction can be stated as follows.

Let A_1OB_1 and A_2OB_2 be two rays (going "from above to below") that refract at the point O. (See Figure 3.1.) The angles α_1 and α_2 formed by the vertical OC and the respective lines A_1O and A_2O are called *incidence angles* (a term with which you should already be familiar). The angles β_1 and β_2 formed by the vertical OD and the respective lines B_1O and B_2O are called *refraction angles*. Snel showed that

$$\frac{\sin \alpha_1}{\sin \beta_1} = \frac{\sin \alpha_2}{\sin \beta_2},$$

that is, the *ratio of the sine of the incidence angle to the sine of the refraction angle is a constant that is independent of the incidence angle.*

Descartes, one of the greatest French thinkers and scholars, arrived at the same law independently of Snel. In the last story in Part One, we will have reason to ponder the question of "whether geniuses err." Well, Descartes was one of the "erring" geniuses. Out of his "errors," scattered over the fields of science, have grown many life-giving shoots.

Descartes deduced the law of refraction from his conceptions of the propagation of light rays. These conceptions have not withstood the test of time, although they led later to the law of conservation of momentum.

Descartes' theory implied that the speed of light is greater in a denser medium, such as water, than in a less dense medium such as air. Many other scientists doubted this. Fermat explained the law of refraction from the opposite assumption, that light moves more slowly in a denser medium.

Fermat and Descartes were both Frenchmen, as well as contemporaries. They often engaged in arguments in the search for scientific truth. This was

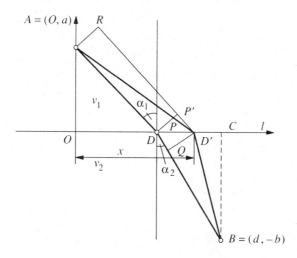

FIGURE 3.2

one such argument. In this case Fermat turned out to be correct. Experiments showed that the denser the medium, the slower the speed of light.

To explain the law of refraction of light, Fermat advanced an extremal principle for optical phenomena. It was later named for him. This principle states that, *in an inhomogeneous medium, light travels from one point to another along the path requiring the shortest time.*

Fermat's principle allows the precise formulation and solution of a minimum problem that leads to the derivation of Snel's law. Specifically, this principle requires the computation of the minimum of the following function of one variable (see Figure 3.2):

$$(1) \qquad f(x) = \frac{\sqrt{a^2 + x^2}}{v_1} + \frac{\sqrt{b^2 + (d - x)^2}}{v_2}.$$

It is worth noting that, at the time that he advanced his extremal principle (approximately 1660), Fermat already had at his disposal an algorithm for finding maxima and minima of functions that was equivalent to setting the derivative equal to zero. The use of derivatives so simplifies the derivation of Snel's law that it now can be carried out by high school students. Fermat himself obtained the required result in a far more elaborate way. It is natural to ask: Why did Fermat not use his algorithm? The answer is very simple: Fermat could apply his method to polynomials—and here he actually anticipated the notion of a derivative—but he did not know how to apply it to radical expressions. That is why the deduction of Snel's law using derivatives was first accomplished by Leibniz, who introduced this concept in the very same work of 1684 in which he laid the foundations of the grandiose edifice of mathematical analysis.

Thus, Fermat deduced Snel's law from his extremal principle, but his solution was very complicated. A far simpler solution, also based on Fermat's principle, was given by Huygens, yet another scientific genius of the seventeenth century and the author of the wave theory of light.

Before reproducing Huygens' solution let's state the problem precisely.

Given two points A and B on either side of a horizontal line l separating two media. It is required to find a point D such that the time it takes for a light ray to traverse the path ADB is a minimum, provided that the velocity of propagation of light is v_1 in the upper medium and v_2 in the lower one (Figure 3.2). Note that (1) is a mathematical reformulation of this problem and that this problem is very similar to Heron's problem.

Huygens' solution. Let D (see Figure 3.2) be a point at which

$$(2) \qquad \frac{\sin \alpha_1}{\sin \alpha_2} = \frac{v_1}{v_2}.$$

We will show that for any other point $D' \neq D$ the time of traversal of the path $AD'B$ is greater than the time of traversal of the path ADB. To this end we erect perpendiculars to the line AD at A and D, respectively. Let P be the point of intersection of AD' and the perpendicular at D. We draw a line through D' parallel to AD and denote its points of intersection with the perpendiculars (to AD) at D and A by P' and R, respectively. Finally, we drop the perpendicular to $D'Q$ from D' to DB. From Figure 3.2 we see that the angles PDD' and $D'DQ$ are respectively equal to α_1 and $\pi/2 - \alpha_2$, respectively. Hence

$$(3) \qquad |D'P'| = |D'D| \sin \alpha_1, \ |DQ| = |DD'| \sin \alpha_2.$$

Now we compare the traversal times along the paths ADB and $AD'B$.

The relation (3) and the inequalities $|AP| > |AD|$, $|D'P| > |D'P'|$, and $|D'B| > |BQ|$ (inclined segments are longer than perpendicular ones) imply that

$$\frac{|AD'|}{v_1} > \frac{|AD| + |P'D'|}{v_1} = \frac{|AD|}{v_1} + |D'D| \frac{\sin \alpha_1}{v_1},$$

$$\frac{|D'B|}{v_2} > \frac{|BQ|}{v_2} = \frac{|DB| - |DQ|}{v_2} = \frac{|DB|}{v_2} - |DD'| \frac{\sin \alpha_2}{v_2}.$$

The latter inequalities and (2) show that

$$\frac{|AD'|}{v_1} + \frac{|D'B|}{v_2} > \frac{|AD|}{v_1} + \frac{|DB|}{v_2}.$$

Thus the refraction point that minimizes the time of traversal of the broken path from A to B is characterized by the fact that the ratio of the sines of the angles of incidence and refraction is equal to v_1/v_2, that is, to a constant. But this is just Snel's law.

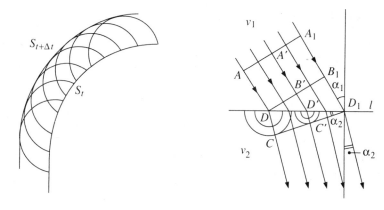

FIGURE 3.3   FIGURE 3.4

What underlies Fermat's principle is the assumption that light is propagated along certain lines. This idea ties in most readily with the corpuscular theory of light that regards light as *a flow of particles*. We owe to Huygens another explanation of the propagation and refraction of light, based on the notion of light as a *wave* whose front moves in time.

A *wavefront* S_t is the set of points that can be reached by light from some source in time t. For example, if at time zero the source is a point and the medium is homogeneous, then at time t the front S_t will be a sphere of radius vt centered at the light source. With increasing distance from the source the spherical wave becomes ever more planelike. Thus if we think of the source as infinitely distant, then the wavefront will be a plane moving uniformly with velocity v.

To determine the motion of a wavefront in more complex cases Huygens used the following rule, now known as "Huygens' principle": *every point of a wavefront S_t itself becomes a secondary source, and in time Δt we obtain a family of wavefronts from all these secondary sources, and the actual wavefront $S_{t+\Delta t}$ at time $t + \Delta t$ is the envelope of this family*—that is, the surface tangent to all secondary wavefronts. (See Figure 3.3.)

Let's use Huygens' principle to deduce Snel's law.

Consider a parallel pencil of light rays falling on a plane boundary separating two homogeneous media. As before, we will suppose that l is horizontal and that the light falls from above. (See Figure 3.4.) We will denote the velocities of propagation of light above and below l by v_1 and v_2 and the angles of incidence and refraction by α_1 and α_2. The wavefront $A_1 A' A$ is moving with velocity v_1 and at a certain moment t reaches the boundary l at the point D. Then D becomes a secondary wave source that propagates in the lower medium with velocity v_2. Light reaches the point D_1 at time $t_1 = t + |B_1 D_1|/v_1 = t + (|DD_1|\sin\alpha_1)/v_1$, and an intermediate point D' on the segment DD_1 at the moment $t' = t + (|DD'|\sin\alpha_1)/v_1$. By time t_1, the spherical wave due to the secondary source D will have radius $r_1 = v_2(t_1 - t) = |DD_1|(v_2/v_1)\sin\alpha_1$ and the wave due to D' will have radius $r' = v_2(t_1 - t') = |DD'|(v_2/v_1)\sin\alpha_1$. Since the angles DD_1C and $D'D_1C$ are equal (their respective sines are $r_1/|DD_1|$ and $r'/|D'D_1|$ and both numbers are equal to $(v_2/v_1)\sin\alpha_1$), the tangents D_1C and D_1C' to these spheres coincide. But D' is an arbitrary point on DD_1. This means that all secondary waves are tangent to the line CD_1 at t_1. The latter line forms with l an angle α_2 such that $\sin\alpha_2 = (v_2/v_1)\sin\alpha_1$. Thus we have again obtained Snel's law.

The idea of a wavefront can also be illustrated with examples that are not derived from optical problems. Consider a traveller who begins to walk at a point A of a rectilinear highway that bounds a meadow. The traveller tries to reach a point B in the meadow as quickly as possible. His speed v in the meadow is half his speed on the highway. If the traveller walks all the time in the meadow, then in a unit of time he can reach any point of a circle of radius v. If he walks all the time on the highway then he will cover the distance $2v$. Suppose he walks partly on the highway and partly in the meadow. Then the set of points that he can reach in a unit of time is a "wavefront" consisting of two segments connected by a circular arc.[1] Now let's touch once more on the subject of extremal principles.

---

[1]The following justification of this claim is due to Prof. E. Barbeau.

To fix ideas, suppose the traveller is at $A(0, 0)$, that he can walk 1 unit per second along the x-axis and $1/2$ unit per second along any other path in the plane.

The most efficient paths to consider are those beginning along the x-axis and then going straight to B.

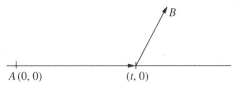

How far can the traveller go in 1 second? Suppose he leaves the x-axis at the point $(t, 0)$, $0 \le t \le 1$ (we consider only the positive quadrant). Then he winds up on the circle

$$C_t : (x - t)^2 + y^2 = \left(\frac{1-t}{2}\right)^2.$$

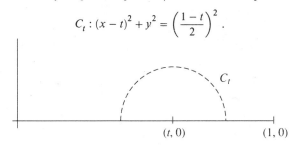

Let us fix a particular $x \in [0, 1]$ and see which circle will maximize y. More specifically,

In this story I have given two derivations of the law of refraction of light. There is a fundamental difference between them. Fermat's approach sheds no light whatsoever on the true essence of the occurring phenomenon. In this approach, one postulates a certain property of the trajectories and shows that it is borne out by experiment. In Huygens' approach, the point of departure is the description of the physical nature of the phenomenon.

This descriptive duality is typical of the natural sciences. While the laws of nature admit of interpretations based on physical models, they are also derivable from extremal principles.

The two approaches described in this story have played a very important role in the history of the calculus of variations and of the whole theory of extremal problems. In fact, every problem in the calculus of variations and in optimal control can be investigated in two ways. One way is to investigate its extremal trajectories (in the manner of Fermat). This leads to the

---

we want to maximize

$$\phi_x(t) = \left(\frac{1-t}{2}\right)^2 - (x-t)^2 = \frac{1}{4}[(1-4x^2) + 2(4x-1)t - 3t^2]$$

$$= \frac{1}{4}\left[\frac{4(1-x)^2}{3} - 3\left(\frac{4x-1}{3} - t\right)^2\right]$$

$$= \left[\frac{1}{3}(1-x)^2 - \frac{3}{4}\left(\frac{4x-1}{3} - t\right)^2\right].$$

If $0 \le x \le \frac{1}{4}$, then $\phi_x(t)$ has its maximum at $t = 0$, so

$$y^2 \le \frac{1}{4} - x^2, \text{ or } x^2 + y^2 \le \frac{1}{4},$$

and $(x, y) \in$ circle with center $(0, 0)$ and radius $\frac{1}{2}$.

If $\frac{1}{4} \le x \le 1$, then $\phi_x(t)$ has its maximum at $t = (4x-1)/3$, so

$$y^2 \le \frac{(1-x)^2}{3}, \text{ or } y \le \frac{1-x}{\sqrt{3}},$$

and (x, y) is under the line $y = (1-x)/\sqrt{3}$.

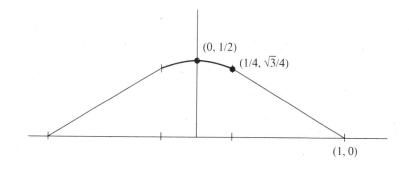

$(0, 1/2)$

$(1/4, \sqrt{3}/4)$

$(1, 0)$

Euler-Lagrange theory (which we will touch upon in the fourteenth story). The other way (the one due to Huygens) is to investigate bundles of extremal trajectories. This procedure leads to analogs of wavefronts, to the theory developed by Hamilton and Jacobi in the nineteenth century, and to the investigation of problems of optimal control by means of the methods of dynamic programming, first developed (relatively recently) by the American scientist Richard Bellman.

4

Maxima and Minima in Geometry

The history of science contains many examples of applications of pure geometry and of its usefulness.

P. Laplace

Archimedes will be remembered when Aeschylus will have been forgotten, for languages die while mathematical ideas do not.

G. H. Hardy

Inexhaustible supplies of precious problems on maxima and minima are hidden in the depths of the oldest mathematical discipline—geometry.

Geometric problems on maxima and minima are found in the works of each of the three greatest mathematicians of antiquity—Euclid, Archimedes, and Apollonius. They were also paid tribute by the most prominent mathematicians of the Renaissance—Viviani, Torricelli, Fermat, and others. Even today, interest in such problems remains high.

1. Euclid's problem. In Euclid's *Elements*, the first scientific monograph and textbook in the history of mankind, which was written in the fourth century B.C., there is just one maximum problem. The following is its modern formulation:

In a given triangle ABC inscribe a parallelogram ADEF (EF∥AB, DE∥AC) of maximal area. (See Figure 4.1 on page 28.)

I will give one of the possible geometric solutions of this problem; it goes back to Euclid's solution in the *Elements*. Specifically, I will prove that what

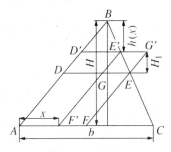

FIGURE 4.1

characterizes the required parallelogram is that D, E, and F are the midpoints of the appropriate sides.

Let $AD'E'F'$ be a parallelogram inscribed in ABC that is different from $ADEF$. Let G' denote the point of intersection of the lines $D'E'$ and EF and G the point of intersection of the lines DE and $E'F'$.

We wish to show that the area of the parallelogram $AD'E'F'$ is less than the area of the parallelogram $ADEF$ by the area of the parallelogram $EG'E'G$. To this end, we drop the altitude from the point B in the triangle ABC and denote its length by H. We denote the length of the side AC by b and the length of the altitude in the triangle $GE'E$ from the point E' by H_1.

In view of the similarity of the triangles $GE'E$ and ABC ($E'G\|AB$ and $GE\|AC$), we have

$$\frac{H_1}{|GE|} = \frac{H}{b} \Leftrightarrow \frac{H_1}{H/2} = \frac{|GE|}{b/2}.$$

From this relation it follows that the area of the parallelogram $D'G'ED$, whose altitude is H_1 and the length of whose side DE is $b/2$, is equal to the area of the parallelogram $EGF'F$, whose altitude is $H/2$ and the length of whose side $F'F$ is $|GE|$. It follows that the area of the parallelogram $ADEF$ is equal to the area of the figure $AD'G'EGF'$ that is greater than the area of $AD'E'F'$ by the area of the parallelogram $GE'G'E$. This completes the solution of the problem.

2. The problem of Archimedes. We have already mentioned that some ancient authors attribute to Archimedes (287–212 B.C.) the proof of the isoperimetric property of the circle and the isoephiphanic property of the sphere. But in the surviving works of Archimedes there is no reference to the isoperimetric problem, and his contribution to its solution is thus far unknown. On the other hand, in his work *On the sphere and cylinder*, Archimedes poses and solves the following problem:

Among all spherical segments with the same spherical area, find the one that encloses the largest volume.

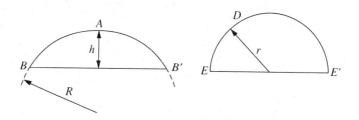

FIGURE 4.2

We will first give a solution which, while entirely based on Archimedes' ideas, nevertheless depends strongly on algebra. We will then give the same solution in the purely geometric language used by its author.

Consider a sphere of radius R and its spherical segment BAB' of height h. (See Figure 4.2.) Together with the segment BAB' we consider the hemisphere EDE' with the same lateral surface. We denote its radius by r. We know that the volume V of the spherical segment is $\pi h^2(R-(h/3))$, its lateral surface area is $2\pi Rh$, the volume $\hat{V}$ of the hemisphere is $(2/3)\pi r^3$, and its lateral surface area is $2\pi r^2$. From the equality of the lateral surface area of the segment and the hemisphere, we have

(1) $$r^2 = Rh.$$

We will prove the inequality

(2) $$(2R - r)r > (2R - h)h \text{ for } h \neq R.$$

We will consider two cases: (a) $h < R$ and (b) $h > R$. In case (a), $r^2 = Rh > h^2 \Rightarrow r > h \Rightarrow R - r < R - h \Rightarrow (2R - r)r = R^2 - (R - r)^2 > R^2 - (R - h)^2 = (2R - h)h$. In case (b),

$$\left.\begin{array}{l} r^2 = Rh < h^2 \\ r^2 = Rh > R^2 \end{array}\right\} \Rightarrow R < r < h$$

$\Rightarrow r - R < h - R \Rightarrow (2R - r)r = R^2 - (R - r)^2 > R^2 - (R - h)^2 = (2R - h)h$. Using (1) and (2) and multiplying by $\pi h/3$ we obtain

(3) $$\frac{\pi h}{3} \cdot 2Rr > \frac{\pi}{3}(3R - h)h^2.$$

Replacing Rh by r^2 in (3) we arrive at the required inequality

$$\hat{V} = \frac{2}{3}\pi r^3 = \frac{\pi h}{3}2Rr > \pi h^2\left(R - \frac{h}{3}\right) = V.$$

Thus a hemisphere whose lateral surface is equal to that of a spherical segment encloses a greater volume than the segment. To quote Archimedes, "of all spherical segments bounded by equal surfaces the largest is a hemisphere."

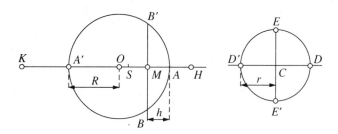

FIGURE 4.3

Of all scientists, the genius of Archimedes, like that of Newton, most likely elicits the greatest admiration. We will re-solve our problem, but this time we will follow Archimedes' thought almost literally (and include in parentheses the relevant algebraic relations).

Archimedes could use neither the language of algebra—whose birth was to come 18 centuries later—nor algebraic computations. His language was that of geometry. Following Archimedes, we lay off on the line $A'A$ (Figure 4.3) a segment $[OH]$ so large that a cone of height HM and base radius MB has the same volume as the spherical segment BAB'. On the segment $[OA']$ produced we lay off the segment $[A'K]$ of length equal to the radius R. From the equality of the volumes of the cone and the segment Archimedes obtains the proportion

$$(4) \qquad \frac{|HM|}{|AM|} = \frac{|KM|}{|A'M|}.$$

We use the familiar formulas for the volume V_K of a cone and V_C of a segment to check this equality:

$$(5) \qquad \begin{aligned} V_K &= \frac{\pi}{3}|HM||MB|^2 = \frac{\pi}{3}|HM||MA'||MA| \\ &= V_C = \frac{\pi}{3}(3R - h)h^2 = \frac{\pi}{3}|KM||AM|^2. \end{aligned}$$

This check uses the fact that the length of the segment $[MB]$ is the geometric mean of the lengths of the segments $[A'M]$ and $[MA]$. The equality (4) follows directly from (5).

The equality of the surface areas of the hemisphere and the segment implies that

$$(6) \qquad |AB| = |ED|.$$

Indeed, $|ED| = r\sqrt{2}$, $|AB|^2 = |AA'||AM|$ (the familiar property of a triangle inscribed in a circle and based on a diameter), so that $\pi|AB|^2 = 2\pi Rh = S_C = \hat{S} = 2\pi r^2 = \pi|ED|^2 \Rightarrow |AB| = |ED|$.

Now Archimedes lays off the segment $[AS]$ equal in length to $[CD]$ and proves the inequality (2): $|A'S||AS| > |A'M||AM|(\Leftrightarrow (2R - r)r > (2R - h)h)$. Archimedes justifies this geometrically: of two rectangles with the same perimeter, the one with the larger area is the one with the greater small side.

In view of the equality of the lateral surface areas of the segment and the hemisphere, we have

$$|AS|^2 = |AM||A'K| \quad (\Leftrightarrow r^2 = Rh).$$

This equality and the preceding inequality yield

$$|AS||AA'| > |KM||AM| \quad (\Leftrightarrow 2Rr > (3R - h)h).$$

Multiplying by $|AM|$ and using (5), we obtain

(7) $$|AS||AA'||AM| > |KM||AM|^2 \quad (\Leftrightarrow 2Rrh > (3R - h)h^2).$$

We showed earlier that

$$|KM||AM|^2 = |HM||MB|^2 \quad [\text{see } (5)],$$
$$|AA'||AM| = |AB|^2 = |ED|^2 \quad [\text{see } (6)].$$

By construction, $|AS| = |CD|$. These equalities and (7) yield

$$\hat{V} = \frac{\pi}{3}|CD||ED|^2 > \frac{\pi}{3}|HM||MB|^2 = V_K = V_C$$
$$(\Leftrightarrow \hat{V} = \frac{2\pi}{3}r^2 > \frac{\pi}{3}(3R - h)h^2 = V_C).$$

This completes the proof.

It may be appropriate to recall that all the formulas we have used (the volume of a cone, a sphere and a spherical segment, the surface area of the sphere and a segment) were first obtained by Archimedes in his work *On the sphere and cylinder*. It is difficult not to agree with Hardy (see epigraph) that Archimedes will be famous as long as mathematics survives. (But I am reluctant to agree with the second half of his sentiment. Aeschylus, too, will remain famous!)

We will postpone discussion of the problem posed and solved by Apollonius until the thirteenth story.

3. Steiner's problem. *In the plane of a triangle, find a point such that the sum of its distances from the vertices of the triangle is minimal.*

This problem was discussed in another formulation in the first story. It, too, has a long history, although not as long as that of Heron's problem or the classical isoperimetric problem. It was included in Viviani's *On maximal and minimal values* (1659), the first work devoted to our subject.

Cavalieri and Torricelli were also interested in this problem. (The solution of this problem, that is, the point where the required minimum is attained, is called the *Torricelli point*; see, for example, Zetel's book [5R].) Coxeter claims that Fermat also studied this problem.

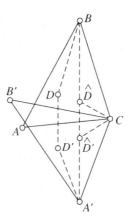

FIGURE 4.4

(Viviani, Cavalieri, and Torricelli were the greatest Italian mathematicians of the seventeenth century. Cavalieri's principle served as a precursor of the integral calculus; Torricelli is known for the discovery of atmospheric pressure. Torricelli and Viviani were students of Galileo. In fact, it was to Viviani that the blind Galileo dictated his *Conversations on mechanics* near the end of his life.)

The interest of so many eminent scientists in so elementary a problem provides yet another confirmation that aesthetic motives often provide the stimulus for creativity.

In the nineteenth century Steiner devoted much attention to this and to a series of similar problems. They are frequently referred to as *Steiner's problems*. We will also use this name.

We will now give the well-known geometric solution of Steiner's problem for triangles whose angles do not exceed $120°$.

Suppose that the angle C in the triangle ABC (Figure 4.4) is $\geq 60°$. We rotate the triangle ABC about C through $60°$ and obtain the triangle $A'B'C$. Let D be any point in triangle ABC and D' be its image under our rotation. Then the sum of lengths $|AD| + |BD| + |CD|$ is equal to the length of the polygonal line $|BD| + |DD'| + |D'A'|$.

Now let $\widehat{D}$ be the Torricelli point, that is, the point from which all the sides of the triangle are seen at an angle of $120°$, and let $\widehat{D}'$ be the image of $\widehat{D}$ under our rotation. It is easy to see that the points $B, \widehat{D}, \widehat{D}'$ and A' are collinear. This means that the Torricelli point is the solution of our problem. We leave it to the reader to show that, if the obtuse angle is greater than $120°$, then its vertex is the solution of the problem.

The fourth and fifth problems in the first story closely resemble Steiner's problem. We leave it to the reader to think through the fourth problem. The answer to this problem can be stated as follows: If the points $A, B, C,$

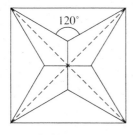

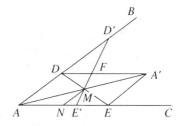

FIGURE 4.5   FIGURE 4.6

and D form a convex quadrilateral, then the required point is the point of intersection of its diagonals; otherwise it is the vertex of the largest angle.

For problem 5, the answer is "yes." For example, in the case of a square there are two extremal nets, represented in Figure 4.5. The sum of the lengths of these nets is less than the sum of the diagonals.

Let's now state two familiar geometric minimum problems.

4. Least area problem. *Given an angle and a point in its interior. To pass a line through the given point that cuts off from the angle a triangle of minimal area.*

We will show that the required line is such that its segment in the interior of the angle is halved by the given point. Such a line is easy to construct. One way is to join the given point M (Figure 4.6) to the vertex A, to lay off on the segment $[AM]$ produced a segment $[MA']$ of length equal to the length of $[AM]$, and to pass through the point A' a line parallel to AC. Let D be the point of intersection of this line and the side AB. It is easy to see that the line joining D to M and intersecting AC at a point E has the required property $|DM| = |ME|$ (the triangles MDA' and MEA are congruent). There are also other constructions of this line.

It remains to show that the line just constructed yields the required minimum. To this end, we pass through M some line $D'E'$. We assume for definiteness that the point E' is to the left of E. Then the area of the triangle $AE'D'$ is equal to the area of the triangle AED minus the area of the triangle EME' plus the area of the triangle MDD'. Let F be the point of intersection of the lines DA' and $D'E'$. Then the triangles EME' and MDF are congruent. Since the latter triangle is contained in the triangle $DD'M$, it follows that the area of the triangle ADE is smaller than the area of the triangle $AD'E'$.

5. Least-perimeter problem. *Given an angle and a point in its interior, pass a line through the given point that cuts off from the angle a triangle of minimal perimeter.*

We will show that the required line DE has the property that the excircle

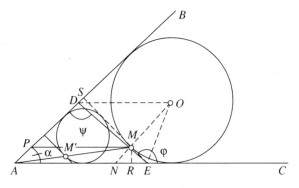

FIGURE 4.7(a)

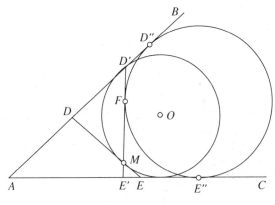

FIGURE 4.7(b)

of the triangle ADE is tangent to the segment $[DE]$ at the point M. It is easy to construct such a line. To this end we inscribe in the angle BAC (Figure 4.7(a)) a circle and denote by M' the point of intersection of the line AM and the circle that is closest to A. Next we draw a tangent to our circle at the point M' and then a line through M parallel to that tangent. This is the required line.

It remains to show that the line just constructed yields the required minimum. To this end we pass through M some line $D'E'$. (See Figure 4.7(b).) Consider the excircle of the triangle $AD'E'$ touching the segment $[D'E']$ at the point F and the sides AB and AC of the angle at the points D'' and E'', respectively. The lengths of the segments $[E'F]$ and $[E'E'']$ are equal, since they are the lengths of tangent segments drawn from the same point. The same is true of the lengths of the segments $[D'F]$ and $[D'D'']$. Hence the perimeter of the triangle $AD'E'$ is equal to the sum of the lengths of the segments $[AD'']$ and $[AE'']$; of course, $|AD''| = |AE''|$. This means that for the perimeter of triangle $AD'E'$ to be minimal the points E'' and D''

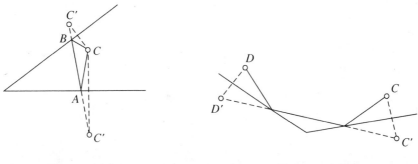

FIGURE 4.8   FIGURE 4.9

must be as close to A as possible. This will take place just when the excircle "leans" on the point M, that is, when the segment of the line through the point M touches that circle at M.

We have not yet paid the "geometric debt" of discussing problems 1 and 2 from the first story. Their solutions are clear from Figures 4.8 and 4.9, respectively. If the constructions shown in these figures are impossible, then, in each case, the required point coincides with the vertex of the relevant angle.

Our topic is inexhaustible. The number of books devoted to it is tremendous. We will mention a few of them. They contain problems to suit any taste. After studying Part Two the reader can try to find analytic solutions to problems. A few books containing geometric extremal problems are Courant and Robbins [3], Coxeter [4], Niven [11], Zetel'[5R], Šarygin [10R], Šklarskiĭ, Čentsov, and Yaglom [12R]. The book of Boltyanskiĭ and Yaglom [13] pays special attention to such problems as well.

We will analyze additional geometric problems in the thirteenth story.

5

Maxima and Minima
in Algebra and in Analysis

Algebra is generous. She often gives more than is
asked of her.

D'Alembert

1. Tartaglia's problem. We will begin our story with a discussion of the
following problem, posed by Niccolo Tartaglia (1500–1557).

To divide the number 8 *into two parts such that the result of multiplying
the product of those parts by their difference is maximal.*

We will attempt to reconstruct the chain of reasoning that led Tartaglia to
the solution of his problem. Before we do so, it will be helpful to say a few
words about the history of his remarkable discovery of the rule for solving
cubic equations by radicals.

The first to solve the equation $x^3+px+q = 0$ (for positive p and negative
q) was Scipione del Ferro (1465?–1526). At that time the only admissible
roots of an equation were its positive roots. Negative roots, and all the more
so complex roots, were ignored.

Del Ferro did not publish his discovery although he did show it to his associates. At that time "mathematical contests" were very popular. (In our own
time this tradition has been revived in a somewhat different form. Groups
rather than individual contestants enter the fray, for example, members of
a boarding school, winners of a school olympiad and its judges, and so on.)
One of those initiated into the secret of the solution of cubic equations decided to use it in order to prevail in such a contest. This competitor would

undoubtedly have succeeded had he not been fated to encounter Niccolo Tartaglia. Tartaglia's task was to solve 30 cubic equations for different values of p and q. At first he was unaware that his opponent knew the secret of the general solution; he discovered this shortly before the deadline for the presentation of the solutions of the problems. Through a tremendous effort, Tartaglia managed, on his own, to find the general method eight days before the deadline. (For details, see Gindikin's book [6].) Like del Ferro before him, Tartaglia obtained the following formula:

$$(1) \qquad x = \sqrt[3]{-\frac{q}{2} + \sqrt{\frac{q^2}{4} + \frac{p^3}{27}}} + \sqrt[3]{-\frac{q}{2} - \sqrt{\frac{q^2}{4} + \frac{p^3}{27}}}.$$

Formula (1) yields an expression for the positive root in del Ferro's case (for $p > 0$ and $q < 0$). But it also yields an expression for a real root in other cases (for example, as we will see, when $p < 0$ and $q > 0$). This formula is usually called the *Cardano formula* in honor of the man who first published it.

Tartaglia did not publish the formula himself, but in a number of his works he announced that he was able to solve problems of various kinds. One of these was the problem given earlier in this story. Without describing the solution, Tartaglia stated the answer in the following form: *Halve the number* 8; *the square of that half augmented by a third of that square will be equal to the square of the difference of the two parts.* In other words, if we denote the required numbers by a and b ($a > b$), then Tartaglia states that $(a-b)^2 = (8\div 2)^2 + (8\div 2)^2 \div 3 = 64/3$, so that $a-b = 8/\sqrt{3} \Rightarrow a = 4 + (4/\sqrt{3})$. We will see that Tartaglia was right.

Following Zeuthen's book [14], we will try to reconstruct Tartaglia's train of thought that led him to the correct answer. Rather than tie ourselves down to the concrete number 8, we will solve the problem in general form. Let S denote the number to be divided. We saw that Tartaglia's answer involves not the numbers a and b but rather their difference, which he almost certainly took as the unknown. If we set $a - b = x$, then $a = (S + x)/2$, $b = (S - x)/2$, so that we are looking for the maximum of the function $f(x) = x(S/2 + x/2)(S/2 - x/2) = (S^2 x - x^3)/4$. Let M denote the maximum in question (for $x \geq 0$). Then we obtain for x the equation

$$(2) \qquad x\left(\frac{S}{2} + \frac{x}{2}\right)\left(\frac{S}{2} - \frac{x}{2}\right) = M \Leftrightarrow x^3 - S^2 x + 4M = 0.$$

Unfortunately, here $p = -S^2 < 0$ and $q = 4M > 0$, so that equation (2) does not have the structure of del Ferro's equation. On the other hand, equation (2) has a noteworthy special feature, namely that in addition to a negative root (that we denote by β) it has *a positive root of multiplicity two*, that is, here the function and its derivative vanish. Figure 5.1 shows that for

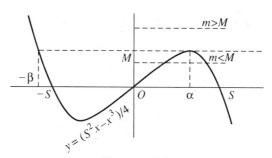

FIGURE 5.1

$m > M$ the equation $x^3 - S^2 x + 4m = 0$ has no positive roots, for $m < M$ it has two such roots, and for $m = M$ it has one positive root. We denote the positive root of equation (2) by α, which is thus the required difference. We can therefore write down the identity

$$x^3 - S^2 x + 4M = (x + \beta)(x - \alpha)^2 = x^3 + (\beta - 2\alpha)x^2 + (\alpha^2 - 2\alpha\beta)x + \alpha^2\beta$$

which implies that $\beta = 2\alpha$, $p = -S^2 = \alpha^2 - 2\alpha\beta = \alpha^2 - 4\alpha^2 = -3\alpha^2$, $q = 4M = \alpha^2\beta = 2\alpha^3$. But then $q^2/4 + p^3/27 = 0 \Leftrightarrow (4M)^2/4 = S^6/27 \Leftrightarrow (2M)^2 = (S^2/3)^3$. It seems that Tartaglia believed that if in (1) $q^2/4 + p^3/27 = 0$, then this formula yields the expression for the negative root:

$$-\beta = 2\sqrt[3]{-\frac{q}{2}} = -2\sqrt[3]{2M} = -2\frac{S}{\sqrt{3}} \Leftrightarrow \beta = \frac{2S}{\sqrt{3}}.$$

Whence

(3) $$\alpha = \frac{\beta}{2} = \frac{S}{\sqrt{3}} \Rightarrow \alpha^2 = \frac{S^2}{3} = \left(\frac{S}{2}\right)^2 + \frac{1}{3}\left(\frac{S}{2}\right)^2.$$

For $S = 8$, we see that in order to find the square of the difference "we must halve the number 8 and augment the square of that half by a third of that square." Thus, the problem is solved.

Many interesting problems on maxima and minima are concealed in various "exact inequalities." We will continue our story with a discussion of what may well be the oldest such inequality.

2. The inequality of the arithmetic-geometric means for two numbers. Let a and b be nonnegative numbers. Their *geometric mean* is the number $\sqrt{ab}$ and their *arithmetic mean* the number $(a + b)/2$. We will show that *for any two nonnegative numbers a and b, we have the inequality*

(1) $$\sqrt{ab} \le \frac{a + b}{2},$$

that is, the geometric mean does not exceed the arithmetic mean. The inequality (1) is *exact* in the sense that in (1) equality is actually attained. This occurs if (and only if) $a = b$.

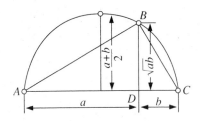

FIGURE 5.2

The inequality (1) conceals various extremal problems. Two such problems are:

   (a) *Find the maximum of the product of two numbers whose sum is constant.*

   (b) *Find the maximal area of a right triangle whose small sides have constant sum.*

One consequence of the inequality (1) is that *among all right triangles with prescribed sum of the small sides, the isosceles triangle has maximal area,* a fact known already to ancient geometers.

Problem (a) is algebraic by content, problem (b) is geometric. When Fermat discovered his method for finding maxima and minima (to be discussed in the eleventh story), he presented it in a private letter to Roberval, a well-known contemporary mathematician, using problem (b) to illustrate his method.

There are many proofs of the inequality (1). We will give two proofs, one of which is algebraic, the other, geometric.

The algebraic proof is based on the following chain of obvious inequalities:

$$0 \le (a-b)^2 \Rightarrow 2ab \le a^2+b^2 \Rightarrow 4ab \le a^2+2ab+b^2 = (a+b)^2 \Rightarrow \sqrt{ab} \le \frac{a+b}{2}.$$

Let's now turn to geometry (Figure 5.2). We take a segment of length $a+b$ ($|AD| = a$, $|DC| = b$) and draw a semicircle with $[AC]$ as diameter. We erect a perpendicular to AC at D and denote by B its point of intersection with the semicircle. In view of the similarity of the triangles ABD and BCD (recall that the angle B subtends a semicircle and is therefore a right angle, so that angle A is equal to angle DBC and angle C is equal to angle ABD), we have

$$\left|\frac{AD}{BD}\right| = \left|\frac{BD}{DC}\right| \Rightarrow |BD| = \sqrt{ab}.$$

If we now keep $[AC]$ fixed (that is, if we are given the sum $a + b$) and vary the point D, then it is clear that the segment BD will attain its maximal length ($= (a + b)/2$) when D coincides with the center of the semicircle. This proves the inequality (1).

3. The inequality of the arithmetic-geometric means (the general case). We will prove the following theorem.

For arbitrary nonnegative numbers $x_1, \ldots, x_n$ *we have the inequality*

(1)
$$\sqrt[n]{x_1 \cdots x_n} \le \frac{x_1 + \cdots + x_n}{n}.$$

The left side of (1) is called the *geometric mean* of the numbers $x_1, \ldots, x_n$ and the right side their *arithmetic mean*. Thus *the geometric mean does not exceed the arithmetic mean*, not only for $n = 2$ but also for arbitrary n. The inequality (1) is *exact*—it becomes an equality only if all the numbers are equal.

There are many proofs of the inequality (1). One of the most beautiful, as well as completely elementary, proofs was formulated by the famous French mathematician A. L. Cauchy.

We begin by using Cauchy's method to prove inequality (1) for $n = 3$. To this end we deduce (1) for $n = 4$ and "descend" to $n = 3$. For $n = 4$ the inequality (1) follows readily if we twice use the established version of (1) for $n = 2$:

(2)
$$x_1 \cdot x_2 \cdot x_3 \cdot x_4 = (x_1 \cdot x_2) \cdot (x_3 \cdot x_4) \le \left(\frac{x_1 + x_2}{2}\right)^2 \left(\frac{x_3 + x_4}{2}\right)^2$$
$$= \left[\left(\frac{x_1 + x_2}{2}\right)\left(\frac{x_3 + x_4}{2}\right)\right]^2 \le \left(\frac{x_1 + x_2 + x_3 + x_4}{4}\right)^4.$$

This proves (1) for $n = 4$. Using (2) we have

$$(x_1 \cdot x_2 \cdot x_3)^{1/3} = [x_1 \cdot x_2 \cdot x_3 \cdot (x_1 \cdot x_2 \cdot x_3)^{1/3}]^{1/4} \le \frac{x_1 + x_2 + x_3 + (x_1 x_2 x_3)^{1/3}}{4}$$

$$\Rightarrow \frac{3}{4}(x_1 \cdot x_2 \cdot x_3)^{1/3} \le \frac{x_1 + x_2 + x_3}{4} \Rightarrow (x_1 \cdot x_2 \cdot x_3)^{1/3} \le \frac{x_1 + x_2 + x_3}{3},$$

which proves (1) for $n = 3$.

Next we will prove (1) in the general case. First, we note that following the approach used earlier to prove the inequality (1) for $n = 4$, it is possible to prove it for $n = 8$, then for $n = 16$, and so on—that is for $n = 2^k$, $k = 2, 3, \ldots$.

Now we will employ the "method of descent" used earlier in going from $n = 4$ to $n = 3$. Suppose the inequality has been proved for $n = m + 1$. We will now prove it for $n = m$. By assumption,

$$(x_1 \cdots x_m)^{1/m} = ((x_1 \cdots x_m)(x_1 \cdots x_m)^{1/m})^{1/(m+1)}$$

$$\le \frac{x_1 + \cdots + x_m + (x_1 \cdots x_m)^{1/m}}{m + 1}.$$

Hence

$$(x_1 \cdots x_m)^{1/m}\left(1 - \frac{1}{m+1}\right) \le \frac{x_1 + \cdots + x_m}{m + 1} \Rightarrow (x_1 \cdots x_m)^{1/m} \le \frac{x_1 + \cdots + x_m}{m}.$$

This completes the proof of inequality (1).

Our proof is but one of many. The well-known book *Inequalities*, by Beckenbach and Bellman, contains twelve proofs of this inequality. Of these, the simplest is probably the one due to Ellers. Following Ellers, we prove by induction that $x_1 \cdots x_n = 1$, $x_i > 0$, implies the inequality $x_1 + \cdots + x_n \geq n$ (from this, the rest follows in an obvious manner). For $n = 1$, this assertion is trivial. Assume that it holds for $n = m$. Let $x_1 \cdots x_{m+1} = 1$. Then there are two numbers (say, x_1 and x_2) such that $x_1 \geq 1$ and $x_2 \leq 1$, that is $(x_1 - 1)(x_2 - 1) \leq 0$ or, equivalently, $x_1 x_2 + 1 \leq x_1 + x_2$. This and the induction assumption imply that $x_1 + \cdots + x_{m+1} \geq 1 + x_1 x_2 + x_3 + \cdots + x_{m+1} \geq 1 + m$, which was to be shown.

The inequality of the arithmetic-geometric means has always been a favorite topic of mathematical clubs. For example, consider the following problem.

In a given sphere, inscribe a cone of maximal volume.

Let R denote the radius of the sphere and r and h the base radius and altitude of the cone, respectively. Then (think this through) the volume V of the cone is equal to $\pi h^2(2R - h)/3$. Using the inequality of the arithmetic-geometric means, we obtain

$$\frac{3V}{4\pi} = \frac{h}{2} \cdot \frac{h}{2}(2R - h) \leq (2R/3)^3,$$

with equality attained for $h/2 = 2R - h \Rightarrow h = (4/3)R$. For this value of the altitude the cone will have maximal volume.

Here are two more problems.

In a given cone inscribe a cylinder of maximal volume.

Given a sheet of tin $a \times b$, cut out equal squares at its corners so that the open box obtained by bending the resulting edges has maximal volume.

Solving such problems by our method is of interest as long as one is not acquainted with differentiation.

4. The inequality of the arithmetic-quadratic means. Let $x_1, \ldots, x_n$ be some numbers. By their *quadratic mean*, we mean the number $[(x_1^2 + \cdots + x_n^2)/n]^{1/2}$. The following theorem is true.

The inequality

(1)
$$\frac{x_1 + \cdots + x_n}{n} \leq \left(\frac{x_1^2 + \cdots + x_n^2}{n}\right)^{1/2}$$

holds for arbitrary numbers $x_1, \ldots, x_n$, that is *the arithmetic mean does not exceed the quadratic mean. The inequality (1) is exact.* It becomes an equality only if all the numbers are equal.

The inequality (1) can also be proved in several different ways. The following proof is probably the simplest. We have

(2) $$0 \le (a-b)^2 \Rightarrow 2ab \le a^2 + b^2.$$

By squaring the arithmetic mean and using the inequality (2) we obtain

$$\left(\frac{x_1 + \cdots + x_n}{n}\right)^2 = \frac{x_1^2 + \cdots + x_n^2 + 2x_1x_2 + 2x_1x_3 + \cdots + 2x_{n-1}x_n}{n^2}$$

$$\le \frac{x_1^2 + \cdots + x_n^2 + (x_1^2 + x_2^2) + (x_1^2 + x_3^2) + \cdots + (x_{n-1}^2 + x_n^2)}{n^2}$$

$$= \frac{n(x_1^2 + \cdots + x_n^2)}{n^2} = \frac{x_1^2 + \cdots + x_n^2}{n},$$

which is what we wished to show.

Juxtaposition of the inequality of arithmetic-geometric means and the (established) inequality (1) shows that *for arbitrary nonnegative numbers* $x_1, \ldots, x_n$ *we have the exact inequality*

$$\sqrt[n]{x_1 \cdots x_n} \le \left(\frac{x_1^2 + \cdots + x_n^2}{n}\right)^{1/2}.$$

In particular, for $n = 2$ we have $\sqrt{x_1 x_2} \le \sqrt{(x_1^2 + x_2^2)/2}$. This inequality can be easily given a geometric interpretation. For example, it directly implies that *of all rectangles inscribed in a circle the square has largest area.* In turn, this problem admits at least two stereometric generalizations. First, *of all rectangular parallelepipeds inscribed in a sphere, find the one of largest volume.* Second, *of all cylinders inscribed in a sphere, find the one of largest volume.* Both of these stereometric problems were investigated by Kepler. (We discuss this further in the next story.) Incidentally, an immediate consequence of the inequality (3) for $n = 3$ is that *of all parallelepipeds inscribed in a sphere, the cube has the largest volume.* (Think this through!)

The planimetric problem of the rectangle of largest area inscribed in a circle will also turn up later. We will refer to it as *Kepler's planimetric problem.*

5. The Cauchy-Bunyakovskiĭ inequality. The following theorem holds: *for arbitrary numbers $a_1, \ldots, a_n, b_1, \ldots, b_n$ we have the inequality*

(1) $$a_1 b_1 + \cdots + a_n b_n \le (a_1^2 + \cdots + a_n^2)^{1/2}(b_1^2 + \cdots + b_n^2)^{1/2}.$$

The inequality (1) is called *the Cauchy-Bunyakovskiĭ inequality.* This inequality is exact: equality is attained for $a_1 = b_1, \ldots, a_n = b_n$. We will prove (1). If $b_1 = \cdots = b_n = 0$, then there is nothing to prove. Suppose that not all b_i are zero. For an arbitrary x, we have

$$(a_1 + xb_1)^2 + \cdots + (a_n + xb_n)^2 = a_1^2 + \cdots + a_n^2 + 2x(a_1 b_1 + \cdots + a_n b_n)$$
$$+ x^2(b_1^2 + \cdots + b_n^2) = ax^2 + 2bx + c,$$

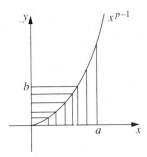

FIGURE 5.3

where we set

$$a = b_1^2 + \cdots + b_n^2, \qquad b = a_1 b_1 + \cdots + a_n b_n, \qquad c = a_1^2 + \cdots + a_n^2.$$

It is clear that $a > 0$ and that for all x we have the inequality

(2) $$ax^2 + 2bx + c \geq 0.$$

But the nonnegative character of the quadratic trinomial in (2) is equivalent to the inequality

(3) $$b^2 - ac \leq 0 \Leftrightarrow (a_1 b_1 + \cdots + a_n b_n)^2 \leq (a_1^2 + \cdots + a_n^2)(b_1^2 + \cdots + b_n^2),$$

which is what we wished to show.

An important generalization of the Cauchy-Bunyakovskiĭ inequality follows.

6. The Hölder inequality. We will show that *the following inequality holds for nonnegative numbers* $a_1, \ldots, a_n, b_1, \ldots, b_n$ *and for* $p > 1$, $p' = p/(p-1)$, $((1/p) + (1/p') = 1)$:

(1) $$a_1 b_1 + \cdots + a_n b_n \leq (a_1^p + \cdots + a_n^p)^{1/p} (b_1^{p'} + \cdots + b_n^{p'})^{7/p'}.$$

The inequality (1) is called *the Hölder inequality.* To prove it, consider the functions $y = x^{p-1}$ and $x = y^{p'-1}$. These functions are mutually inverse (check this). Choose two positive numbers a and b. Then

$$\int_0^a x^{p-1} dx = \frac{a^p}{p}, \quad \int_0^b y^{p'-1} dy = \frac{b^{p'}}{p'}.$$

Now look at Figure 5.3. The quantity a^p/p is the area of the vertically cross-hatched curvilinear triangle, and $b^{p'}/p'$ is the area of the horizontally cross-hatched curvilinear triangle. It is easy to verify that, regardless of the disposition of a and b, the sum of the areas of these two triangles is not less than the area of the rectangle with sides a and b. Also, equality is possible only if $a^{p-1} = b$. This means that for any two nonnegative numbers a and

b, we have the inequality

(2)
$$ab \le \frac{a^p}{p} + \frac{b^{p'}}{p'}.$$

Now let $a_1, \ldots, a_n$ and $b_1, \ldots, b_n$ be arbitrary nonnegative numbers. If, say, $b_1 = \cdots = b_n = 0$, then the inequality (1) holds. We can therefore assume that $A = (a_1^p + \cdots + a_n^p)^{1/p} \ne 0$ and $B = (b_1^{p'} + \cdots + b_n^{p'})^{1/p'} \ne 0$. Put $x_k = a_k/A$, $y_k = b_k/B$. In view of (2), we have

$$x_k y_k \le \frac{a_k^p}{pA^p} + \frac{b_k^{p'}}{p'B^{p'}}, \qquad k = 1, \ldots, n.$$

Adding these inequalities and bearing in mind that $1/p + 1/p' = 1$ and that $a_1^p + \cdots + a_n^p = A^p$, $b_1^{p'} + \cdots + b_n^{p'} = B^{p'}$, we end up with the required inequality

$$x_1 y_1 + \cdots + x_n y_n \le 1 \Rightarrow a_1 b_1 + \cdots + a_n b_n \le AB \Rightarrow a_1 b_1 + \cdots + a_n b_n$$
$$\le (a_1^p + \cdots + a_n^p)^{1/p} (b_1^{p'} + \cdots + b_n^{p'})^{1/p'}.$$

The topic of exact inequalities is unusually extensive. Many books and papers deal with such inequalities. One of the best known is the book [7] by Hardy, Littlewood, and Polya. This topic is also well covered in the popular literature.

As a rule, using the general methods that we will talk about in the second half of this book, it is possible to prove many exact inequalities without difficulty. But there are exceptions. Here are two problems that are easy to state, but proving their respective extremal properties is not, I think, a simple matter. The reader should try to solve them. He may hit on some simple solutions.

PROBLEM 1. *Find the least value of the sum of the fourth powers of an odd number of quantities $x_1, \ldots, x_{2l+1}$ given that their sum and the sum of their cubes are both zero and the sum of their squares is one.*

PROBLEM 2. *A hundred positive numbers $x_1, \ldots, x_{100}$ satisfy the conditions $x_1^2 + \cdots + x_{100}^2 > 10000$, $x_1 + \cdots + x_{100} < 300$. Show that there are among them three numbers whose sum is greater than 100.*

6

Kepler's Problem

Near a maximum the decrements on both sides are
in the beginning only imperceptible.

J. Kepler

Goethe wrote that "when you confront Kepler's life story with what he
became and what he achieved, you are at once joyfully astounded and convinced that true genius is bound to overcome all obstacles." This story depicts
one of the most radiant, noble, and exalted geniuses that has ever existed.
Since I am not in a position to illuminate this remarkable personality with
a measure of thoroughness, I must refer the refer to two books on Kepler.[1]
In his book [8R], Predtečenskiĭ describes with rare beauty Kepler's moral
eminence. In his recent book *Johann Kepler* [3R], Belyĭ describes in detail
Kepler's scientific progress and his genius. Belyĭ's book also includes an extensive bibliography.

It seems to have been Kepler's fate to be spared no trial. He endured
poverty, privation, sickness, the death of loved ones, upheavals, and exile.
And yet, when we read him, we are invariably conscious of his thankfulness
to fate for the gift of joy and happiness, the joy of labor, and the pursuit of
truth. This is how he rhapsodized about his third law of planetary motions:

> I yield freely to the sacred frenzy; I dare frankly to confess
> that I have stolen the golden vessels of the Egyptians to build
> a tabernacle for my God far from the bounds of Egypt. If
> you pardon me, I shall rejoice; if you reproach me, I shall

---

[1]Two English books on Kepler are: J. Banville, *Kepler: A novel*, Secker and Warburg, London,
1981, and A. Koestler, *The sleepwalkers*, Hutchinson, London, 1959; Penguin, New York, 1964.
(A.S.)

endure. The die is cast, and I am writing the book—to be
read either now or by posterity, it matters not. It can wait a
century for a reader, as God himself has waited six thousand
years for a witness.

Kepler seems to have derived much the same joy from the discoveries of
others. This is how he writes to Galileo when expressing delight at the latter's
discovery of the satellites of Jupiter:

I stayed home, did nothing, and thought of you, dear and
famous Galileo when I suddenly learned of your discovery
of four planets with the aid of the telescope... I could not
think without extreme agitation that in this way our ancient
argument was resolved... I may seem somewhat daring if I
so readily trust your claims unsupported by any of my own
tests. But why should I not believe the most learned of mathematicians whose correctness is confirmed by the very mode
of his reasoning.

Kepler referred to Snel as the Apollonius of his time. When he ran into
difficulties in solving geometric problems, he addressed Snel thus:

Produce for us, oh Snel, glory of the geometers of our time,
solutions of this and other problems that are now required.

Kepler was utterly convinced that any persons pursuing the truth will be
happy to learn of its discovery. He addressed such people in these words:

In some places it is necessary to dwell at length, so that ...
learned men would know what to profit by and what to enjoy.

Predtečenskiĭ writes of Kepler:

He is forever unaffected and true to himself. Conceit and
ambition are foreign to his lofty mind. He sought neither
honors nor praise. He never claims to be superior to scholars that are now virtually unknown, and all his life referred
with profound respect to Maestlin, whose sole distinction is
that he had the good fortune of having Kepler for a student
... Tycho Brahe ... was his chief antagonist, for he rejected the Copernican theory so zealously advocated by Kepler. We know that the relations between the two great men
were marred by many unpleasant incidents. And yet Kepler
invariably praises Tycho, gives him his due, and makes no
attempt to diminish his merits ... Here, and at all times,
Kepler shows himself to be a champion of the truth. Sad to
say, this is all too rare ... in our own time.

Kepler had every reason to say of himself, "I am used to telling the truth everywhere and at all times."

The theme of genius is not well developed in Russian literature. It was not dealt with by either Tolstoy or Dostoevsky. Pushkin is an exception. The word "genius" is articulated in *Mozart and Salieri*, but most of the time Pushkin uses the more inclusive term "Poet."

Artlessness, loyalty in friendship, the creative urge, the ability to delight in all that is beautiful and, of course, an inability to do evil—all these are characteristics of genius that Pushkin bestowed on his Mozart and that were also markedly present in Kepler. Like the Poet, he was invariably guided by the motto: "Follow the free mind wherever it leads." Just as "the Poet alone chooses the subjects of his poems," so too Kepler chose the subjects of his researches.

This story is devoted to one such subject.

In his book, *New solid geometry of wine barrels*, Kepler describes an event in his life that occurred in the fall of 1613:

> In December of last year ... I brought home a new wife
> at a time when Austria, having brought in a bumper crop
> of noble grapes, distributed its riches ... The shore in Linz
> was heaped with wine barrels that sold at a reasonable price
> ... That is why a number of barrels were brought to my
> house and placed in a row, and four days later the salesman
> came and measured all the tubs, without distinction, without
> paying attention to the shape, without any thought or com-
> putation. Namely the copper point of a ruler was pushed
> through the filling hole of a barrel, across the heel of each of
> the wooden disks which we refer to simply as bottoms, and
> as soon as the length to the point at the top of one board
> disk was the same as the length to the point at the bottom of
> the other, the salesman stated the number of amphoras con-
> tained in the barrel after merely noting the number on the
> ruler at the spot where the length in question ended. I was
> astonished ...

Kepler thought it strange that by means of a single measurement (see Figure 6.1 on page 50, taken from Kepler's book), one could determine the volumes of barrels of different shape. He goes on:

> Like a bridegroom, I thought it proper to take up a new sub-
> ject of mathematical studies and to investigate the geometric
> laws of a measurement so useful in housekeeping, and to clar-
> ify its basis if such exists.

In order to clarify a basis of this kind Kepler had to lay the foundations of

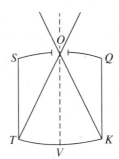

FIGURE 6.1

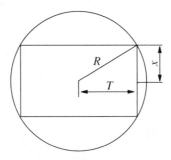

FIGURE 6.2

differential and integral calculus, as well as advance new ideas for the solution of maximum and minimum problems.

The key result in the book *New solid geometry of wine barrels* is Theorem V [Part Two]: "Of all cylinders with the same diagonal, the largest and most capacious is that in which the ratio of the base diameter to the height is $\sqrt{2}$." In other words, this theorem provides the solution of the following problem: *Inscribe in a given sphere a cylinder of maximal volume.* The corresponding problem in the plane is *to inscribe in a given circle a rectangle of maximal area.* Hereafter we will call the first of these problems *Kepler's problem* and the second *Kepler's planimetric problem.*

To begin, we solve Kepler's problem by a method that Tartaglia would have employed (had he posed it). Let R be the radius of the sphere. We denote by x half the height of the cylinder. (See Figure 6.2.) Then the base radius is $\sqrt{R^2 - x^2}$ and the volume of the cylinder is $2\pi(R^2 - x^2)x$. We recall that in Tartaglia's case we had $(S^2 - x^2)x/4$. The formula in the preceding story yields the maximum value $\hat{x} = R/\sqrt{3}$ and

$$\hat{r} = \sqrt{R^2 - \frac{R^2}{3}} = R\sqrt{\frac{2}{3}}.$$

This implies that in the extremal cylinder the ratio of the base diameter to

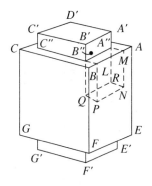

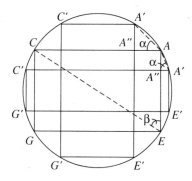

FIGURE 6.3   FIGURE 6.4

the height is $\sqrt{2}$. This coincides with Kepler's result.

In solving this problem Kepler could have used his idea of the insensitivity of the variation of a function near its maximum (see the epigraph for this story). But he ignored this possibility and provided a purely geometric solution.

Kepler reduced the problem of the most capacious cylinder to the solution of the following maximum problem: *Of all rectangular parallelepipeds with a square base inscribed in a sphere the one with largest volume is the cube.*[2] This result is proved in Theorem IV of Part Two of Kepler's book.

For brevity's sake, Kepler called a rectangular parallelepiped with square base a *post*, and we will do likewise. We distinguish two cases: (a) the post is higher than the cube; and (b) the post is lower than the cube.

Let's look first into case (a). (See Figures 6.3 and 6.4.) Consider the cube $ABCDEFGH$ and the "post" $A'B'C'D'E'F'G'H'$ inscribed in the same sphere (the points D, H, and H' are invisible in Figure 6.3). We compare their volumes. Two parallelepipeds with square bases protrude from the cube, namely $A'B'C'D'A''B''C''D''$ above it and one of equal volume below it.

But far more can be "subtracted" from the cube. This is easy to see: At each side of the square $A''B''C''D''$ there is a parallelepiped that borders on the post whose base is a square congruent to $A''B''C''D''$. We denote one of these parallelepipeds by $A''B''QRMNPL$. The volume of these four parallelepipeds alone exceeds that of the protruding parts of the post. In fact, the volume of the protruding parts is equal to $2|A''B''|^2|A''A'|$ and the volume of the bordering parallelepipeds is $4|A''B''|^2|A''M|$. But $|A''M| = |A''A|/\sqrt{2}$.

Now we consider the triangle $A'AA''$ (Figure 6.4). The angle $\alpha = \widehat{A'AA''}$ subtends the arc $A'C$ and the angle $\beta = \widehat{AEC}$ subtends the larger arc

---

[2]This is a special case of the problem discussed in the previous story.

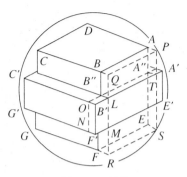

FIGURE 6.5

AC. This means that $\alpha < \beta$, and thus $|A''A'|/|A''A| = \tan\alpha < \tan\beta = |CA|/|AE| = \sqrt{2}$. It follows that

$$2|A''B''|^2|A''A'| < 2|A''B''|^2|A''A|\sqrt{2}$$
$$= 2|A''B''|^2\sqrt{2}\sqrt{2}|A''M| = 4|A''B''|^2|A''M|.$$

This proves the required inequality of volumes.

It remains to consider the case (b). (See Figure 6.5.) Again, let the cube $ABCDEFGH$ and the post $A'B'C'D'E'F'G'H'$ be inscribed in the same sphere (the points H, D', and H' are invisible in Figure 6.5). We compare their volumes. The combined volume of the two parallelepipeds with square base that protrude from the post is $2|AB|^2|AA''|$. The volume of the part of the post that protrudes from the cube is less than this combined volume. Kepler proves this in the following manner.

Let us, says Kepler, attach to each lateral face of the cube a parallelepiped, or *panel*, of the same thickness as the protruding part of the post. (One such panel, namely $ABFEPQRS$, is shown in Figure 6.5). Their combined volume is $4|AB|^2|AP| = 4|AB|^2|A''A'|/\sqrt{2}$. Again, by considering the triangles $AA'A''$ and AEC (in Figure 6.4), we show that angle α is greater than angle β, so that $|AA''|/|A''A'| > \sqrt{2}$. But, as Kepler correctly observes, even if we attach the four panels to the cube, parts of the post remain exposed. They are four "gaping" parallelepipeds at the edges of the post (one such is $B''LB'OF''MF'N$ shown in Figure 6.5 (F'' is invisible)). Each of these is part of a little post erected at one of the edges AE, BF, CG, and DH of the cube. The volume of each of these four little posts is $|AB||AP|^2$. Now the panels we applied earlier to the lateral faces of the cube stick out beyond the height of the post by eight little panels (one of which is $BQPATA''B'L$ in Figure 6.5). The volume of each of these eight little panels is $|AB||AP||AA''|$. The inequality $2|AA''| > 2\sqrt{2}|A''A'| = 4|AP|$ implies that the volume of the four little posts at the edges of the cube is less than the volume of the eight little panels. Thus the volume that protrudes from the post is $2|AB|^2|AA''| > 4|AB|^2|AP|$, while the volume that protrudes from the cube is less than $4|AB|^2|AP| - 8|AB||AP||AA''| + 4|AB||AP|^2 < 4|AB|^2|AP|$. It follows that in the transition from cube to post, the cube loses more than it gains. This completes the solution of the auxiliary problem. (We recall that in the previous story we investigated a somewhat more general problem by algebraic means.)

The rest of the proof is very simple. In every cylinder one can inscribe a post and the ratio of their volumes—in this order—is constant and equal to $\pi/2$ (check this). This means that the cylinder of largest volume inscribed in a sphere is the one in which we can inscribe a cube. And in such a cylinder the ratio of base diameter to height is $\sqrt{2}$.

After proving this theorem Kepler wrote:

> From this it is clear that, when making a barrel, Austrian barrelmakers, as if guided by common and geometric sense, take as the radius of a bottom a third of the length of a stave. When this is done, the cylinder constructed in the mind between two bottoms will consist of two halves, each of which will be close to the conditions of theorem V and will thus have maximal capacity even if one deviated somewhat from the exact rules during the making of the barrel, because figures closed to the optimal change their capacity very little ... This is so because near a maximum the decrements on both sides are in the beginning only imperceptible.

Kepler's concluding words contain the fundamental algorithm for finding extrema that was later shaped into an exact theorem. First described (for polynomials) by Fermat (1629) and then, in general form, by Newton and Leibniz, this algorithm was later called "Fermat's theorem." A great deal of interesting information about Kepler's problem can also be found in M. B. Balk's article, "The secret of the old barrelmaker" (*Kvant*, 1986, 8, p. 14, in Russian).

7

The Brachistochrone

If one considers motions with the same initial and terminal points then, the shortest distance between them being a straight line, one might think that the motion along it needs least time. It turns out that this is not so.

Galileo Galilei

The profound significance of well-posed problems for the advancement of mathematical science is undeniable.

D. Hilbert

Acta Eruditorum, the first scientific journal, began publication in 1682. In the June 1696 issue of this journal, there appeared a note by the famous Swiss scholar Johann Bernoulli with the intriguing title, "A new problem that mathematicians are invited to solve."

It is often the case that the statement of a new problem attracts the attention of many eminent scholars. By competing with one another they create powerful methods for the solution of problems that later offer great service to science. This was the case with Johann Bernoulli's problem. Its author stated it as follows:

Let two points A and B (Figure 7.1 on page 56) be given in a vertical plane. Find the curve that a point M, moving on a path AMB must follow such that, starting from A, it reaches B in the shortest time under its own gravity.

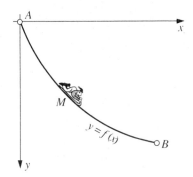

FIGURE 7.1

When posing his problem, Bernoulli made no mention of Galileo. How unfair! All modern natural science "issued" from Galileo. Not only did he discover the fundamental laws of mechanics, but Galileo also was the first to *put questions* to Nature. The present stage of the development of science began when Galileo ascended the tower of Pisa to ask Nature about the laws of falling bodies.

Galileo experimented with inclined planes and, apparently, also with circular chutes. We quote from *Discourses on mechanics*, his life's main work:

> Experience shows that bodies falling down circular arcs corresponding to chords inclined with respect to the horizon ...
> perform motions that also take equal time intervals, shorter than those for motions along the chords.

Of Galileo's two assertions on motions along circular arcs, only one is true: a motion along an arc is faster than one along a chord. The claim about the equality of time intervals is only approximately correct, and, as it turned out later, this fact is intimately related to Bernoulli's problem.

Be that as it may, Galileo's assertion, and his assertion that serves as an epigraph for this story, both must face Bernoulli's question: which curve corresponds to the *shortest* time interval, that is, which curve is the *brachistochrone* (Greek for quickest)? Many authors upbraid Galileo for having mistakenly claimed that a circular arc is a brachistochrone. In the *Discourses* Galileo returned on a number of occasions to the topic of comparing motion on a circle with motion on a chord, but in no place can his words be interpreted as the claim that among all curves joining two points, motion along a circular arc is shortest. However, it is conceivable that we have overlooked some such pronouncement of his.

Many mathematicians responded to Johann Bernoulli's "invitation." One of the first to solve the brachistochrone problem was Leibniz, to whom the problem appealed and who called it "splendid." Next Jakob Bernoulli (Johann's brother) and l'Hospital announced their success. And, of course, Jo-

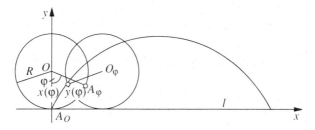

FIGURE 7.2

hann Bernoulli himself had a solution. All of these scholars made significant contributions to the emerging new school, that of mathematical analysis. There was also an anonymous solution identified by experts as provided
by Newton (who later admitted that it took him 12 hours of uninterrupted
reflection to arrive at a solution). *Ex ungue leonem* (tell a lion by his claw)
was Johann Bernoulli's comment on Newton's solution.

All of these mathematicians arrived at the same conclusion. *The brachistochrone is the cycloid*. At this point it is appropriate to examine this remarkable curve.

A cycloid is the path described by a point on a circle that rolls without
sliding on a straight line. Let's derive its equation.

Let l be a horizontal line and let a circle with radius R and center O
roll along l. Suppose that at time zero the point to be observed is the point
of contact of the circle and the line l. We denote it by A_0. Consider the
rectangular coordinate system with A_0 as origin and l as x-axis (Figure 7.2).
We wish to determine the position of A_0 following a clockwise rotation of
the circle through φ. To this end we mark on the original circle the point A_φ
such that the angle $A_\varphi O A_0$ is φ. When the circle will have turned through
an angle φ, A_φ will be the new point of contact with the line l. Since the
length of the arc from A_0 to A_φ is $R\varphi$, this will be the abscissa of the new
position of the center of the circle. The new position of the point A_0 will
be such that the "new version" of the angle $A_0 O A_\varphi$ is again φ. Hence, the
coordinates $(x(\varphi_0), y(\varphi_0))$ of A_0 will be

$$x(\varphi) = R(\varphi - \sin \varphi), \qquad y(\varphi) = R(1 - \cos \varphi).$$

This, then, is the equation of the cycloid that passes through the origin for
$\varphi = 0$. In general, we have an additional parameter

(1) $$x(\varphi) = R(\varphi - \sin \varphi) + C_1, \qquad y(\varphi) = R(1 - \cos \varphi).$$

What is so remarkable about this curve? How did it arise?

The cycloid first turned up in the works of Galileo as an illustrative example. He called it cycloid, meaning "circle-related." The curve was soon
rediscovered in France (by Mersenne, Roberval, Descartes, and Pascal) and

named a *roulette* or a *trochoid*. The first marvel that involved it was that it
became a kind of firing range for trying out new forms of the weapons that
subsequently entered the arsenal of mathematical analysis.

Ancient mathematics bequeathed very few curves to subsequent generations. The main curves studied in antiquity were the circle and the conics,
that is, the ellipse, the hyperbola, and the parabola, curves which turned up
in the works of Appolonius. We should also mention the quadratrix, the
cissoid, the conchoid, and the spiral. Luckily, the first laws of mechanics
did not go beyond this supply of curves; the planets move along ellipses, and
thrown objects describe parabolic arcs.

The best mathematicians of the seventeenth century (including, in addition
to those named previously, Viviani, Torricelli, and some others) perfected
their new methods of investigation on the cycloid; they obtained tangents to
it, determined areas under it, computed the length of its arcs, and so on.

Then came the second marvel. The cycloid became the first "nonancient"
curve connected with the laws of nature. It turned out that the cycloid, and
not—as Galileo wrote—the circle, has the property that a body that glides
along it without friction oscillates with a period unaffected by its initial position. This *tautochrone* (equal-time) property of the cycloid was discovered by
Huygens, and produced a long-lasting sensation. Huygens himself wrote that
"The most desirable fruit, a kind of high point of Galileo's teaching about
falling bodies, is my discovery of the property of the cycloid." This was the
second appearance of the cycloid in a completely new context.

Let's now solve the problem. Recall that there were five solutions due,
respectively, to Johann Bernoulli, Leibniz, Jakob Bernoulli, l'Hospital, and
Newton. All of them were of great interest. Leibniz used a method which
was further developed by Euler (its essence can be surmised from Leibniz's
letter to Johann Bernoulli quoted in the sequel). Nowadays, the Liebniz-
Euler method is one of the basic methods for the solution of problems on
maxima and minima and is known as *the direct method of the calculus of
variations*. Jakob Bernoulli based his solution on Huygens' principle and
thus took another step toward the creation of the Hamilton-Jacobi theory
(mentioned briefly in the third story). But the most popular solution has
been that found by the author of the problem. It has been reproduced in
countless books, and we too will reproduce it here.

First, introduce in the plane a rectangular coordinate system with horizontal x-axis and *downward-directed* y-axis. We place the point A at the origin
(Figure 7.1). Let $y = f(x)$ be the equation of the curve (chute) joining the
point A to the point B with coordinates (a, b). We must now determine
the time it takes a body M of mass m to fall (without friction) from A
to B along the chute $f(x)$. From mechanics, we know Galileo's law which
asserts that the velocity of a body at a point with coordinates $(x, f(x))$, in
a frictionless motion under gravity, is independent of the form of the curve

joining A to $(x, f(x))$ and depends solely on the ordinate $f(x)$. In the words of Johann Bernoulli, "The velocities of falling weighted bodies are to each other as the square roots of the traversed altitudes."

In fact, the kinetic energy of the body at $(x, f(x))$ is $mv^2/2$ and is equal to the difference $mgf(x)$ of the potential energies. In sum, the velocity at $(x, f(x))$ is $\sqrt{2gf(x)}$, where g is the acceleration due to gravity. Next we consider the portion of the path between the points $(x, f(x))$ and $(x + dx, f(x + dx))$, where dx is a small increment of the abscissa. The length ds of this portion of the path is approximately equal to $\sqrt{dx^2 + (f(x + dx) - f(x))^2}$. Using the approximate equality $f(x + dx) - f(x) \approx f'(x)\, dx$, we have $ds = \sqrt{1 + (f'(x))^2}\, dx$. On a small portion of the path, it is reasonable to suppose the velocity constant and equal to $\sqrt{2gf(x)}$. This means that the time required to traverse it is approximately equal to $dt \approx \dfrac{\sqrt{1 + (f'(x))^2}}{\sqrt{2gf(x)}}\, dx$, and that the total time T of the motion from A to B is given by the integral

$$(2) \qquad T = \int_0^a \frac{\sqrt{1 + (f'(x))^2}}{\sqrt{2gf(x)}}\, dx.$$

This leads to the following analytic form of the brachistochrone problem. To find the minimum of the integral $\int_0^a \left(\sqrt{1 + (f'(x))^2} / \sqrt{2gf(x)} \right) dx$ over all functions f with $f(0) = 0$, $f(a) = b$.

We have expressed our problem in mathematical language—more specifically, in the language of the integral calculus. This procedure is called the *formalization* of the problem. (More will be said about this matter in the tenth story.) In the Soviet Union, the elements of integral calculus are now taught in the last (tenth) year of school. Since this book is intended for a larger audience of high school students and not just for those who have completed the tenth class, we will carry out our derivation without the aid of the calculus, in the spirit of seventeenth-century mathematics. In this connection we recall that the basic notions of analysis were introduced just 12 years before Johann Bernoulli's paper, and that the period of "rigor" was as yet in the distant future.

We divide the segment $[0, b]$ of the ordinate axis into n parts by means of the points $0 = y_0;\ y_1, y_2, \ldots, y_n = b$ and find abscissas x_i such that $f(x_1) = y_1$, $f(x_2) = y_2, \ldots, f(x_{n-1}) = y_{n-1}$, $x_n = a$. We join the points (x_i, y_i) and (x_{i+1}, y_{i+1}), $i = 0, 1, \ldots, n - 1$, by line segments. In this way, in addition to the function $y = f(x)$, we obtain a broken line L_n that also joins A and B. (See Figure 7.3 on page 60.) The greater the number n, the closer this broken line approximates the function $y = f(x)$. Also, the gliding time of the body M along this broken line will be close to its gliding time along the chute given by the function $y = f(x)$.

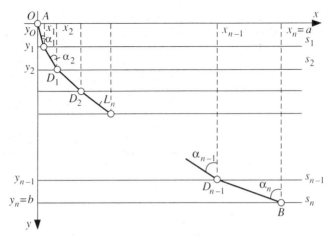

FIGURE 7.3

We may suppose the velocity on the ith segment constant and equal to $\sqrt{2gy_{i+1}}$, $i = 0, 1, \ldots, n - 1$. With this assumption, the exact time of traversal of the n-link broken line is

$$(2_n) \qquad T_n = \frac{\sqrt{y_1^2 + x_1^2}}{\sqrt{2gy_1}} + \frac{\sqrt{(y_2 - y_1)^2 + (x_2 - x_1)^2}}{\sqrt{2gy_2}}$$
$$+ \cdots + \frac{\sqrt{(y_n - y_{n-1})^2 + (x_n - x_{n-1})^2}}{\sqrt{2gy_n}},$$

and this time is to be minimized.

The limit of T_n as $n \to \infty$ is the required time of motion of the body M along the curve $y = f(x)$. This limit coincides with the previously introduced integral (2). But we will not solve the problem formulated earlier by means of integral calculus. Rather, following Johann Bernoulli, we will solve the approximate "discrete" problem corresponding to (2_n). Its formulation follows.

Given two points A and B with respective coordinates $(0, 0)$ and (a, b). On lines $l_1, \ldots, l_n$, parallel to the x-axis and having respective ordinates $y_1, \ldots, y_n$, find points $D_1 = (x_1, y_1), \ldots, D_{n-1} = (x_{n-1}, y_{n-1})$ such that the sum T_n is minimal (assuming that $x_0 = y_0 = 0$, $x_n = a$, and $y_n = b$).

To solve this problem, Johann Bernoulli applied a remarkable method that has strongly influenced all of the subsequent history of the natural sciences. It is this method that we want to discuss next.

The optical-mechanical analogy. Let's go back to the third story. In that story we formulated and solved the problem of refraction of light. Now we will thoroughly investigate its content and compare it with the problem (2_n) for $n = 2$. In both cases we are given two points and a horizontal line and must find on that line a point so as to minimize the sum of "weighted" lengths.

The difference in this case is that, whereas in the third story the velocities v_1 and v_2 were arbitrary and we essentially minimized the function

$$(3) \qquad \frac{\sqrt{y_1^2 + x_1^2}}{v_1} + \frac{\sqrt{(a - x_1)^2 + (b - y_1)^2}}{v_2}$$

(for $A = (0, 0)$, $B = (a, b)$, and l given by $y = y_1$), here we are dealing with a special case, namely we are required to minimize the function

$$(3') \qquad \frac{\sqrt{y_1^2 + x_1^2}}{\sqrt{2gy_1}} + \frac{\sqrt{(a - x_1)^2 + (b - y_1)^2}}{\sqrt{2gb}}.$$

Leibniz arrived at the same conclusion and then went his own way. He gave a wonderful description of his method in his letter of June 16, 1696, to Bernoulli:

> My method is somewhat different from yours but leads to the same result; to respond, as is just, to your openness with the same, here it is in a few words: upon replacing the curve with a polygon with infinitely many sides I see that of all possible cases (the curve) of quickest descent will be obtained if we choose on the broken line any three points, or vertices, A, C and B, and C is such that of all possible points located on the horizontal line l it alone yields the quickest path from A to B. In this way the task is reduced to the solution of an easy problem: given two points A and B and a horizontal line l between them; find on that line the point C such that the path ACB is quickest.

(We have changed Leibniz's notation slightly, to fit that used in this text; he has B instead of C, C instead of B, and DE instead of l.)

Johann Bernoulli proceeded differently. Imagine a nonhomogeneous optical medium consisting of n homogeneous layers $s_1, \ldots, s_n$ (see Figure 7.3), say, n sheets of different kinds of glass. Let the speed of propagation of light in the sheet s_1 be $\sqrt{2gy_1}$, in the sheet s_2, $\sqrt{2gy_2}, \ldots$, and in the sheet s_n, $\sqrt{2gy_n}$. What will be the time for the propagation of light if it is forced to move along the broken line L_n? Of course, the answer is T_n as given by formula (2_n). In other words, light, according to Fermat's principle discussed in the third story, "solves" the very problem (2_n) (if we stipulate that the velocity of propagation of light in the ith layer is $\sqrt{2gy_i}$.

Beginning with a mechanical problem and following Johann Bernoulli we have arrived at an optical problem. This is the first application of the optical-mechanical analogy that was to yield so many discoveries in the works of Hamilton, Jacobi, de Broglie, and many others.

Now we will deal with Johann Bernoulli's solution of the brachistochrone problem. We apply Snel's law (discussed in the third story) to the optical variant of the problem (2_n). Let α_i be the incidence angle in the ith layer. By Snel's law,

$$(4) \qquad \frac{\sin \alpha_1}{\sqrt{2gy_1}} = \frac{\sin \alpha_2}{\sqrt{2gy_2}} = \cdots = \frac{\sin \alpha_n}{\sqrt{2gy_n}} = \text{constant}.$$

If we allow the layers to grow thinner and more numerous then, in the limit, (4) yields

$$(5) \qquad \frac{\sin \alpha(x)}{\sqrt{2gf(x)}} = \text{constant},$$

where $\alpha(x)$ is the angle between the tangent to the curve $y = f(x)$ at the point $(x, f(x))$ and the y-axis. The tangent of this angle is $1/f'(x)$. It follows that $f'(x) = \tan(\pi/2 - \alpha(x)) = \cos \alpha(x)/\sin \alpha(x)$, so that

$$\sin \alpha(x) = \frac{1}{\sqrt{1 + (f'(x))^2}}.$$

This relation and (5) imply that

$$\sqrt{1 + (f'(x))^2}\sqrt{f(x)} = D,$$

where D is some constant. In other words, the function must satisfy the differential equation

$$(6) \qquad y' = \sqrt{\frac{C - y}{y}}.$$

In Bernoulli's time it was known that (6) is the differential equation of a cycloid.

For those with a measure of experience in integration, we integrate the equation (6).

$$y' = \sqrt{\frac{C - y}{y}} \Leftrightarrow \frac{\sqrt{y}\, dy}{\sqrt{C - y}} = dx.$$

We make the substitution $y = C \sin^2(t/2) = C(1 - \cos t)/2$. Then

$$dx = \frac{\sqrt{C} \sin(t/2) d(C \sin^2(t/2))}{\sqrt{C} \cos(t/2)} = C \sin^2(t/2)\, dt = C(1 - \cos t)\, dt/2.$$

Integrating the latter relation we obtain

$$(7) \qquad x = \frac{C}{2}(t - \sin t) + C_1, \qquad y = \frac{C}{2}(1 - \cos t).$$

If we replace $C/2$ by R and t by φ in (7), then we obtain the equation (1) for a cycloid.

These formulas yield a simple prescription for constructing the cycloid that solves Johann Bernoulli's problem. Note that all the cycloids (7) are similar as well as convex. For example, take any cycloid in (7) that has the point

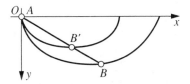

FIGURE 7.4

$A = (0, 0)$ as its left vertex. (See Figure 7.4.) Denote by B' the point of intersection of the line AB and the selected cycloid. To obtain the required cycloid we need only apply to the selected cycloid the similarity with center $(0, 0)$ and coefficient $|AB|/|AB'|$.

The solution of the brachistochrone problem gave its author the tremendous joy of original discovery. Bernoulli said:

> I cannot refrain from expressing once more my amazement at the noted unexpected identity of Huygen's tautochrone and our brachistochrone ... Nature always operates in the simplest manner. Thus in this case it renders two different services by means of one and the same curve.

Johann Bernoulli's method made possible the solution of a number of other remarkable problems in optics, mechanics, and geometry. Let's look at two such problems.

PROBLEM 1. *What are the trajectories of light rays in an atmosphere for which the velocity of propagation of light is proportional to the altitude?*

This problem was investigated by l'Hospital, author of the first textbook of analysis in history. (Alas, it was recently discovered that the essential part of this textbook is an edited version of the lectures given to l'Hospital by the very same Johann Bernoulli.) L'Hospital succeeded in integrating the equation and in answering the question. After approximately 200 years, it turned out that the solution to Problem 1 is directly related to Lobačevskian geometry—the trajectories of light rays coincide with *Lobačevskian straight lines in the Poincaré model.*

PROBLEM 2. *Find the minimal surface of revolution.*

Here we must add a few words to clarify the problem. Let $y = f(x)$ be a nonnegative function passing through two points (x_0, y_0) and (x_1, y_1) in the plane. Rotating it about the x-axis yields *a surface of revolution.* Its area is given by the integral

$$S = 2\pi \int_{x_0}^{x_1} f(x)\sqrt{1 + (f'(x))^2}\, dx.$$

The task is to find the curve $y = f(x)$ that minimizes the area S.

The problem of the minimal surface was solved by Johann Bernoulli and by Leibniz. The reader should try to solve these two problems himself. I will discuss these two problems further in the fourteenth story.

The brachistochrone problem was destined to play an important role in mathematical analysis. In fact, it turned out to be the first of a series of problems underlying the formulation of the calculus of variations.

Shortly before the brachistochrone problem emerged, Newton considered a similar problem. Newton's problem is the subject of the next story.

8

Newton's Aerodynamical Problem

This book (Newton's *Principia*) will forever remain a monument to the profundity of a genius.

P. Laplace

Do geniuses make mistakes? Usually, anyone who asks such a question expects an affirmative answer. There is an element of comfort in the knowledge that even geniuses can err. Sometimes their mistakes elicit comments marked by less than benevolent enthusiasm. Newton too was on the receiving end of this kind of enthusiasm. Thus, in a book on optimal control, you can read that

> Newton formulated a variational problem dealing with a solid of revolution that offers least resistance to a gas. He assumed a physically absurd law of resistance. As a result, the problem he posed has no solution (the more serrated the profile, the smaller the resistance)... Had Newton's arguments been even approximately correct, we would not need expensive wind tunnel experiments today.

What is the problem in question, and how justified is the quoted criticism? This story will look at just these issues.

Newton's *Mathematical principles of natural philosophy* appeared in 1687. No other work in the mathematical literature can be compared with it. A description of the system of the universe discovered by Newton, it contains the kind of discovery that can be made only once. Lagrange called it "the greatest work of the human mind," and Laplace lauded it as "a monument to the profundity of genius."

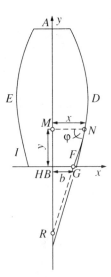

FIGURE 8.1

The book deals with the basic laws of mechanics discovered by Newton, as well as the laws of planetary motion and other fundamental facts. However, part of the text is devoted to many special problems.

While discussing the resistance offered to material bodies by the medium through which they move, Newton tossed off, as if in passing, the following phrase:

> If the figure $DNFG$ is such a curve, that if, from any point thereof, as N, the perpendicular NM let fall on the axis AB, and from the given point G there be drawn the right line GR parallel to a right line touching the figure in N, and cutting the axis produced in R, MN becomes to GR as GR^3 to $4BR \cdot GB^2$, the solid described by the revolution of this figure about its axis AB, moving in the before-mentioned rare medium from A towards B, will be less resisted than any other circular solid whatsoever, described of the same length and breadth.
>
> *Principia*, v. 1 pp. 334–334

This phrase attracted the attention of Newton's contemporaries some nine years later, in 1696, when Johann Bernoulli posed his brachistochrone problem (discussed in our seventh story).

While the brachistochrone problem elicited universal admiration, Newton's problem fared like poor, neglected Cinderella. As a rule, it was brought up—if at all—as an instance of the error of genius. But just as Cinderella's day came, so did the day of Newton's problem.

Let's try to understand Newton's idea.

When constructing shells, torpedos, or rockets, one tries to shape them so as to minimize the resistance they will meet while in motion. Newton writes: "figures may be compared together as to their resistance; and those may be found which are most apt to continue their motions in resisting mediums." There is just so much symmetry we can give to a boat or a plane, for example. But when it comes to the head of a shell, torpedo, or rocket, it stands to reason tht their cross-sections must be circular; in other words, they must take the shape of a *solid of revolution*. But which solid of revolution? A sphere, a cone, a spindle, or yet another circular shape? Such questions cannot be answered without computations, without solving a maximum or minimum problem. Newton posed just such a problem (and gave its solution in the phrase quoted above).

The very first approximation to the problem follows.

PROBLEM. *Find the solid of revolution of given length and width that is subject to least resistance while moving in some medium.*

A few clarifying remarks are in order. The terms used in the formulation of the problem (length, width, motion, and medium) require precise description.

Newton assumes that the front and back of the solid of revolution are the same, that is, that the solid is symmetric with respect to the plane passing through the midpoint of the axis of revolution and perpendicular to it. Thus the length of the solid is its length measured along its axis of rotation, and its width is the radius of its middle section. This being so, it is clear that it suffices to consider half the solid. This is what we will do in the sequel.

Now we consider the motion. Following Newton, we will assume that the body is moving with constant velocity v.

Finally, let's look at the medium. This is the most delicate and central issue. Newton calls it a "rare" medium. He thinks of a rare medium as "consisting of equal particles freely disposed at equal distances from each other." Each of the motionless particles has a fixed mass m and is a perfectly elastic ball. Newton assumed that the body itself is also perfectly elastic. This means that when one of the small balls collides with the moving body, it recoils in accordance with the law that "the angle of incidence is equal to the angle of reflection."

We could now pose the general problem and embark on its solution. However, we will proceed differently. Our first step will be the solution of a simpler problem. (Newton himself solves this simpler problem first.)

PROBLEM OF THE FRUSTUM OF A CONE. *Determine the dimensions of the frustum of a cone subject to least resistance when moving in a rare medium given its base and altitude.*

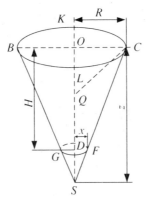

FIGURE 8.2

We will determine the resistance offered by the frustum of a cone moving in a rare medium.

Let H be the altitude of the frustum and R the radius of the upper base. (See Figure 8.2; the boldface part of this figure appears in the *Principia*).

Newton writes:

> For since the action of the medium upon the body is the same... whether the body move in a quiescent medium, or whether the particles of the medium impinge with the same velocity upon the quiescent body, let us consider the body as if it were quiescent.
>
> *Principia*, vol. 1, p. 331

We will do the same, that is, we will assume that the frustum is at rest and that the medium "comes against it" from below with velocity v.

The part of the surface of the frustum that is subjected to collisions with the particles of the medium is its lower base and side. First, we compute the resistance to which the lower base is subject. Let its radius be x. The particles that impinge on this base during a unit of time were originally in a cylinder whose base is the same as the lower base of the frustum and whose altitude is v. The volume V_0 of this cylinder is $\pi x^2 v$. Let ρ be the density of the medium and m the mass of a single particle. Then the number of particles that impinge on the lower base of the frustum during a unit of time is $N_0 = \frac{\rho}{m} V_0 = \frac{\rho}{m} \pi x^2 v$. Upon collision with the lower base, each particle reverses its velocity, so that its momentum increases by $-2mv$. By Newton's third law, the frustum gains an opposite increase of momentum. This means that its total gain of momentum—due to N_0 particles—is $N_0 \cdot 2mv = 2\pi \rho x^2 v^2$. Analogous considerations apply to the side of the frustum. The side collides with particles contained in a hollow cylinder whose volume V_1 is $\pi(R^2 - x^2)v$. The number of particles impinging on the side of the frustum

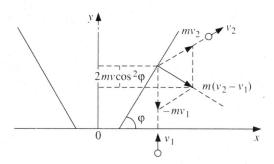

FIGURE 8.3

is $N_1 = (\rho/m)V_1 = (\rho/m)\pi(R^2 - x^2)v$. As for the change in momentum, we note that the gain in momentum of a particle reflected from the side of the frustum is $m(v_2 - v_1)$. (See Figure 8.3.) This vector must be projected on the y-axis. It is easy to see from Figure 8.3 that this projection is $-2mv \cos^2 \varphi$, where φ is the angle between a generator of the frustum and the plane of its lower base. It follows that the total gain of momentum due to the particles impinging on the side of the frustum is

$$N_1 \cdot 2mv \cos^2 \varphi = 2\pi\rho(R^2 - x^2)v^2 \cos^2 \varphi.$$

We will also note (for future use) the following result. Consider the frustum of a cone obtained by revolving about the y-axis a segment $[AB]$ whose endpoints have abscissas a and b and that makes an angle φ with the x-axis. This side of the frustum is subject to a force of resistance given by

(1) $$F = K(b^2 - a^2) \cos^2 \varphi, \qquad K = 2\pi\rho v^2.$$

It follows that the total resistance offered by the frustum is given by

$$F(x) = K\left[x^2 + (R^2 - x^2)\cos^2 \varphi\right], \qquad K = 2\pi\rho v^2.$$

The expression for $\cos \varphi$ in terms of x is

$$\cos \varphi = (R - x)/\sqrt{(R - x)^2 + H^2}.$$

Since the constant K has no effect on the behavior of maxima and minima, we can ignore it.

We can formalize the problem of the frustum of a cone as follows:

PROBLEM 1. *Find the minimum of the function*

(2) $$f(x) = x^2 + (R^2 - x^2)\frac{(R - x)^2}{(R - x)^2 + H^2}$$

for $0 \le x \le R$.

As a rule, a particular translation of a problem into the language of mathematics is not unique. In the present case we can also give the following alternative description of our problem.

In Figure 8.2, extend the segment CF to the point S of its intersection with the y-axis and take the length of the segment OS as the new variable z. The similarity of the triangles SOC and SDF implies that $x/R = (z - H)/z$, where $R - x = RH/z$ and $x = R(z - H)/z$. Substituting these expressions in (2) we obtain a new expression for the force of resistance (divided by K) in terms of the new variable.

$$(3) \qquad g(z) = R^2 \frac{(z - H)^2 + R^2}{z^2 + R^2}.$$

Here z varies from H to infinity. We have obtained the following formulation of the problem of the frustum of a cone (we can ignore the multiplicative constant R^2).

PROBLEM $1'$. *Find the minimum of the function*

$$(2') \qquad h(z) = \frac{(z - H)^2 + R^2}{z^2 + R^2}$$

subject to the condition $z \geq H$.

This problem is so simple that we can solve it without the use of the differential calculus. Our answer will take the same form as Newton's answer.

Let m be the least value of the function h, so that $h(z) \geq m$ for $z \geq H$. Also, $m \leq h(H) = R^2/(R^2 + H^2) < 1$. We make the obvious transformations

$$(4) \qquad \begin{aligned} &h(z) \geq m && \text{for } z \geq H \\ \Leftrightarrow\ &(z - H)^2 + R^2 - mz^2 - mR^2 \geq 0 && \text{for } z \geq H \\ \Leftrightarrow\ &z^2(1 - m) - 2zH + H^2 + R^2(1 - m) \geq 0 && \text{for } z \geq H. \end{aligned}$$

If it turns out that there is an $m < 1$ for which the inequality (4) holds for all z, and if for this m we can find a $\hat{z} \geq H$ that turns this inequality to an equality, then our problem will have been solved. We will try to find the required m and $\hat{z}$.

If $az^2 + 2bz + c \geq 0$ for all z and $a\hat{z}^2 + 2b\hat{z} + c = 0$, then it follows that $D = b^2 - ac = 0$ and $\hat{z} = -b/a$ (why?). In our case, $a = (1 - m)$, $b = -H$, $c = H^2 + R^2(1 - m)$. The equality $D = 0$ yields

$$D = H^2 - (1 - m)(H^2 + R^2(1 - m)) \Rightarrow R^2 m^2 - (2R^2 + H^2)m + R^2 = 0,$$

whence

$$m = \frac{2R^2 + H^2 - H\sqrt{4R^2 + H^2}}{2R^2}.$$

We have discarded the " $+$ " sign because m must be less than 1. Further,

$$(5) \qquad \hat{z} = -b/a = H/(1 - m) = \sqrt{R^2 + (H/2)^2} + H/2 (> H).$$

We have also found $m < 1$ and $\hat{z} \geq H$ such that the discriminant of the equation $z^2(1 - m) - 2zH + H^2 + R^2(1 - m) = 0$ is zero, and so the equation has just one root $\hat{z} \geq H$, that is, $h(z) \geq m$ and $h(\hat{z}) = m$. Problem $1'$ is solved.

What follows is Newton's answer to the problem of the frustum of a cone:

> As if upon the circular base $CKBL$ [Figure 8.2] from the center O, with the radius OC, and the altitude OD, one would construct a frustum $CBGF$ of a cone, which should meet with less resistance than any other frustum constructed with the same base and altitude and going towards D in the direction of its axis: bisect the altitude OD in Q, and produce OQ to S, so that QS may be equal to QC, and S will be the vertex of the cone whose frustum is sought.
>
> *Principia*, vol. 1, p. 333

It is easy to see that the content of Newton's geometric answer is the same as that of the algebraic answer given by formula (5).

The cone subject to least resistance is indeed blunt rather than pointed. Newton goes even further, letting the altitude H tend to zero. Then $\hat{z}$ tends to R, and the angle at the base of the cone tends to $45°$. These facts prompt Newton to suggest replacing an oval body with a blunt one; more specifically, to place in front a circular disk that makes an angle of $135°$ with the adjoining surface. "*This Proposition*," says Newton, "*I conceive may be of use in the building of ships.*"

Newton's problem for a broken line with two links. Let's now take the next-to-last step in the solution of Newton's problem. We propose to solve the special case of this problem for broken lines with two links for which the breaks are located on the line $x = R/2$. In other words, we imitate Leibniz's method.

The precise statement of the present problem calls for finding a point A on the line $x = R/2$ such that the surface obtained by revolving the broken line OAB (where B has coordinates (R, H)) about the y-axis (Figure 8.4) is subject to least resistance in Newton's rare medium.

Suppose that the segment $[OA]$ makes an angle φ_0 with the x-axis and $[OB]$ an angle φ. Let y denote the ordinate of A. Using formula (1) we find that the solid of revolution generated by the broken two-link line OAB is subject to a force of

$$F = K \left[(R/2)^2 \cos^2 \varphi_0 + 3(R/2)^2 \cos^2 \varphi_1 \right], \qquad K = 2\pi \rho v^2,$$

where

$$\cos \varphi_0 = \frac{R}{2\sqrt{y^2 + (R/2)^2}}, \qquad \cos \varphi_1 = \frac{R}{2\sqrt{(H - y)^2 + (R/2)^2}}.$$

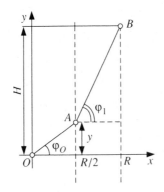

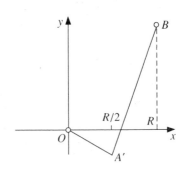

FIGURE 8.4   FIGURE 8.5

This means that we must minimize the function

$$g_1(y) = \frac{1}{a^2 + y^2} + \frac{3}{a^2 + (H - y)^2} \, ;$$

here we have again neglected the multiplier $KR^4/2^4$ and have set $R/2 = a$.

It is easy to see that if $|y|$ increases without bounds, then the function g_1 stays positive and tends to zero. This means that the minimum of g_1 is zero, but this minimum is never attained (a similar situation occurs and will be analyzed in detail in the eleventh story). At first glance it may appear that our problem is pointless. However, physical considerations show that we must restrict y to the interval $[0, H]$. This important point calls for a clarification.

If $y < 0$, then we obtain a broken line $OA'B$, as shown in Figure 8.5. This broken line gives rise to a solid of revolution with a crater. Some of the small balls that constitute the rare medium would be reflected a number of times from the surface of the crater. The body would offer greater resistance, and that resistance would be governed by a different law. Thus the physics of the problem disallows negative values of y. Similarly, values of y greater than H must be ruled out. In other words, the implicit assumption is the *monotonicity of the revolved curve*. This leads to the following formalization of the problem of the broken line with two links.

PROBLEM 2. *Find the minimum of the function*

$$g_1(y) = \frac{1}{a^2 + y^2} + \frac{3}{a^2 + (H - y)^2}$$

subject to the restriction $0 \le y \le H$.

The solution of Problem 2—which is easy to obtain by means of the differential calculus—can be stated as follows:

(a) There exists a $\delta > 0$, such that for $0 \le H \le \delta$ the minimum is attained at 0;

(b) For $H > \delta$ the minimum is attained at an interior point of the interval $[0, H]$. Also,

$$R/4 \tan \varphi_0 \cos^4 \varphi_0 = 3\frac{R}{4} \tan \varphi_1 \cos^4 \varphi_1.$$

This means that for sufficiently small values of H the solid of revolution generated by a two-link broken line is blunt (that is, is the frustum of a cone) and for larger values of H it is pointed (that is, it consists of a cone and the side of the frustum of a cone).

Now let's look at the last intermediate step in solving Newton's problem.

Newton's problem for a broken line with n links. Consider the vertical lines $l_1, l_2, \ldots$ where respective equations are

$$x = \frac{1}{n}, \qquad x = \frac{2}{n}, \ldots$$

(see Figure 8.6), and, if $R \neq k/n$ for some k, the vertical line $x = R$. We consider the totality of monotonically increasing broken lines with breaks on the lines l_k joining $(0, 0)$ to (R, H). We will try to find that one of our curves which, when revolved about the y-axis, generates a solid of least resistance in a rare medium. For an admissible curve, let y_k denote the ordinate of the vertex on the line l_k, and let φ_k denote the angle between the link joining the vertices on the lines l_k and l_{k+1} and the x-axis. Assume, for the sake of simplicity, that $R = N/n$. By Formula (1), the body generated by revolving the curve about the y-axis is subject, in a rare medium, to the

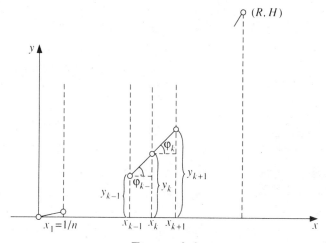

FIGURE 8.6

force of resistance

$$F = K \left\{ \frac{1}{n^2} \cos^2 \varphi_0 + \left[\left(\frac{2}{n} \right)^2 - \left(\frac{1}{n} \right)^2 \right] \cos^2 \varphi_1 + \cdots \right.$$

$$\left. + \left[\left(\frac{N}{n} \right)^2 - \left(\frac{N-1}{n} \right)^2 \right] \cos^2 \varphi_{N-1} \right\}$$

(7)

$$= \frac{K}{n^2} [\cos^2 \varphi_0 + 3 \cos^2 \varphi_1 + \cdots + (2N-1) \cos^2 \varphi_{N-1}],$$

$$K = 2\pi \rho v^2, \qquad \cos \varphi_k = \frac{1}{n} \frac{1}{\sqrt{1/n^2 + (y_{k+1} - y_k)^2}}.$$

Thus Newton's problem for an N-link broken line is the problem of minimizing the force F for all choices of the $(N-1)$-tuples $(y_1, \ldots, y_{N-1})$ with $0 \le y_1 \le y_2 \le \cdots \le y_{N-1} \le H$.

To solve this problem, we'll make use of the experience acquired in solving Problem 2, for, in a sense, the problem for an N-link broken line is made up of $N-1$ problems like Problem 2. Indeed, assume that we have been able to solve the problem for an N-link broken line and that $(\hat{y}_1, \ldots, \hat{y}_{N-1})$ are the ordinates of the vertices of the minimal broken line. Let us fix all but its kth vertex and choose its value so as to minimize the value of the resistance. This is the same as solving a variant of Problem 2 where it is required to find the minimum of the function

$$g_k(y) = \frac{1}{n^2} [(2k-1) \cos^2 \varphi_{k-1} + (2k+1) \cos^2 \varphi_k]$$

$$= \frac{1}{n^4} \left[\frac{2k-1}{\frac{1}{n^2} + (y - \hat{y}_{k-1})^2} + \frac{2k+1}{\frac{1}{n^2} + (\hat{y}_{k+1} - y)^2} \right]$$

for $\hat{y}_{k-1} \le y \le \hat{y}_{k+1}$. The function $g_k(y)$ is made up of the $(k-1)$th and the kth term of the sum (7) for the force F, except that the ordinate of the kth vertex is not fixed and is denoted by y (as usual, we have left out the multiplier K). It is clear that the solution of this problem is $\hat{y}_k$.

The problem of minimizing the function g_k is similar to Problem 2. It follows that, *mutatis mutandis*, the conclusions that hold for Problem 2 apply here. Specifically,

(a) there exists a number $\delta_k > 0$ such that for $\hat{y}_{k+1} - \hat{y}_{k-1} \le \delta_k$ the minimum is attained for $\hat{y}_k = \hat{y}_{k-1}$;

(b) for $\hat{y}_{k+1} - \hat{y}_{k-1} > \delta_k$ we have the equality

(8) $$\left(k - \frac{1}{2} \right) \tan \varphi_{k-1} \cos^4 \varphi_{k-1} = \left(k + \frac{1}{2} \right) \tan \varphi_k \cos^4 \varphi_k.$$

Together, (a) and (b) imply that the extremal broken line follows for a time the x-axis, $\hat{y}_1 = \cdots = \hat{y}_s = 0$, then goes up as per rule (8). The latter means that the value of φ_{s+1} is obtained from the equality

$$Y = \left(s - \frac{1}{2}\right) \tan\varphi_s \cos^4\varphi_s = \left(s + \frac{1}{2}\right) \tan\varphi_{s+1} \cos^4\varphi_{s+1},$$

the value of φ_{s+2} from the equality

$$Y = \left(s + \frac{1}{2}\right) \tan\varphi_{s+1} \cos^4\varphi_{s+1} = \left(s + \frac{2}{3}\right) \tan\varphi_{s+2} \cos^4\varphi_{s+2},$$

and so on.

The force of resistance for an N-link broken line can be written directly in terms of the ordinates of the vertices of the broken line. To this end, we rewrite (7) as

$$\cos\varphi_k = \frac{n^{-1}}{\sqrt{1/n^2 + (y_{k+1} - y_k)^2}}.$$

By making rather obvious changes, we obtain for F the expression

$$F = 2K\left(\sum_{k=1}^{N-1} \frac{k - \frac{1}{2}}{n} \frac{1}{1 + \left(\frac{\Delta y_k}{\Delta x}\right)^2} \Delta x\right);$$

here $\Delta y_k = y_{k+1} - y_k$, $\Delta x = \frac{1}{n}$.

Bearing in mind the form of integral sums and the approximate equality $(\Delta y_k/\Delta x_k) \approx f'(x_k)$, we see that as N tends to infinity, our sum tends to the integral

$$(8') \qquad\qquad F = 2K \int_0^R \frac{x\,dx}{1 + (f'(x))^2}$$

Solution of Newton's problem. It can be shown that as N increases, the minimal N-link broken line tends to the minimal curve that is the solution of Newton's problem. It follows (recall our description of the minimal N-link broken line) that the minimal curve is constructed as follows: *at first the extremal function $f(x)$ coincides with the x-axis, that is, $f(x) = 0$ for $0 \le x \le a$, and then its values go up along some curve (Newton's curve)* subject to the condition

$$(9) \qquad\qquad x \tan\varphi(x) \cos^4\varphi(x) = \text{constant}.$$

Here $\varphi(x)$ denotes the angle between the tangent to the graph of the function $y = f(x)$ at $(x, f(x))$ and the x-axis. The equality (9) is the limiting form of the equalities (8) associated with our solution of the N-link broken line problem.

We will now rewrite (9). In view of the geometric sense of the derivative, we have $\tan\varphi(x) = f'(x)$. Hence $\cos\varphi(x) = [1 + (f'(x))^2]^{-1/2}$. But then

$$(10) \qquad\qquad \frac{x f'(x)}{[1 + (f'(x))^2]^2} = \text{constant}.$$

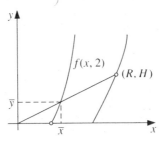

FIGURE 8.7

Equation (10) is the *differential equation of Newton's curve.*

A final remark. At the moment when the horizontal curve goes over into Newton's curve, the derivative of Newton's curve is 1 (a datum that "... may be of use in the building of ships"). The differential equation (10) and the condition that the derivative of the curve at the break is 1 make it possible to find the equation of Newton's curve. In the fourteenth story, we will integrate this equation. In the meantime we will simply present the formulas for the (x, y) coordinates of Newton's curve $y = f(x, c)$:

$$(11) \quad x = c\left(\frac{1}{u} + 2u + u^3\right), \qquad y = c\left(\log\frac{1}{u} + u^2 + \frac{3}{4}u^4\right) - \frac{7c}{4}.$$

Here c is determined by the condition $f(c) = H$.

A few questions remain.

1. When all is said and done, how is Newton's problem solved? How do we construct the curve of given length and width which, when revolved about the y-axis yields the surface of the body subject to least resistance in a rare medium?

The equations in (11) show that all presumed minimal curves depend on one parameter and are similar to each other. This means that to solve Newton's problem we can proceed as in the case of the brachistochrone.

We draw the curve for, say, $c = 2$. (See Figure 8.7.) It can be shown that $f(x, 2)$ intersects any line $y = kx$, $k > 0$, just once.

We draw the line $y = (H/R)x$ joining the origin to the point (R, H). Let $(\overline{x}, \overline{y})$ be the point of intersection of this line with the graph of the function $y = f(x, 2)$. Set $\overline{c} = H/\overline{y}$. Then $y = f(x, \overline{c})$ is the curve that passes through the required point (R, H). Of all the curves in the family (11), this curve is the only one having this property. As such, it is the solution of Newton's problem.

2. What is the meaning of Newton's mysterious phrase quoted in the beginning of this story? What connection is there between this phrase and our solution of Newton's problem?

Consider Figure 8.1 (which contains some of the required problems). We set $|MN| = x$, $|MB| = y$, and $|BG| = b$ and denote by φ the angle between the segment $[MN]$ and the tangent to the curve at N. φ is equal to the angle HGR. Also, $\tan \varphi = f'(x)$. This implies that

$$|BR|/|BG| = \tan \varphi \Rightarrow |BR| = bf'(x).$$

Hence

$$|GR|^2 = |BG|^2 + |BR|^2 = b^2(1 + f'(x)^2).$$

Now Newton's proportion

$$|MN| : |GR| = |GR|^3 : (4|BR| \cdot |GB|^2)$$

yields

(12) $$\frac{x}{b\sqrt{1 + (f'(x))^2}} = \frac{b^3(1 + (f'(x)^2)^{3/2}}{4bf'(x)b^2} \Leftrightarrow \frac{xf'(x)}{(1 + (f'(x))^2)^2} = \frac{b}{4}.$$

This is the differential equation for Newton's curve. (See equations (10) and (12).)

Newton also anticipated the bluntness of the solid of revolution and the break at G (where, we recall, the angle is $135°$). Also, while answering the first question, we showed that, given the length and width, the differential equation (12) and the condition at the break determine the required curve uniquely. We can therefore say that *Newton solved the aerodynamical problem completely.*

3. Why was Newton's curve relegated to the role of the unfortunate Cinderella for 300 years? Why was Newton's idea not fully understood for so long?

I said earlier that the brachistochrone problem opened a new era—the era of the calculus of variations. This subject experienced intense growth for almost two centuries. As I mentioned earlier, it is only recently that we have realized that many technical, and for the most part cosmological, problems of current interest cannot be treated with the methods of the calculus of variations. Instead, the need for a new step forward became apparent. The new theory—which incorporates the calculus of variations—is known as the theory of optimization, or optimal control. This theory has made it possible to solve problems of the new type. *Newton's problem belongs to the category of problems of optimal control.* Within the framework of this theory Newton's problem has a natural and standard solution. On the other hand, this problem has no natural and standard solution within the framework of the calculus of variations. *Thus this problem put Newton* 300 *years ahead of his time!*

4. What is a rare medium? Does it exist? Is it not absurd that a solid of revolution subject to least resistance should be flat in the front? Who has ever dreamed of a torpedo or rocket with a flat head?

Indeed, neither water, nor the surrounding air, nor the usual liquid or gaseous media exhibits the properties of Newton's rare medium. This means that Newton's solution is useless for the construction of motor boats launches, or ocean liners. But in the mid-fifties, when the era of supersonic and high-altitude flying machines began, Newton's physical assumptions and his aerodynamical problem became scientific frontline news. "Up there" the medium is "rare." *Newton's remark about blunted cones turned out to be "of use in the building of ships" of the supersonic and high-altitude variety.*

5. Do geniuses err? In a paper in the journal *Quantum* (1982, 5, pp. 11–18), I posed this question without supplying an answer. But when Andrei Kolmogorov happened to be in the editorial office and someone read him the article, he demanded that the question be answered in the affirmative. Well, maybe.

Newton's problem is certainly a remarkable mathematical event. For 250 years it seemed likely that it had no physical basis and its solution was absurd. But the "mistake" of a genius turned out to have been an insight.

In a word, hasty judgments are sometimes just that. *It can happen that the thought of a genius, which we regard as a mistake carries within it the imprint of truth—a truth clear to him but hidden as yet from us.*

PART TWO

Methods of Solution of Extremal Problems

We must make it our goal to find a method of solution of all problems...by means of a single simple method.

D'Alembert

9

What is a Function?

This general concept requires that by a function of x one should mean the number given for every x and one that varies gradually together with x. The value of a function may be given by an analytic expression, or by a condition that gives the means of testing all numbers and of selecting one of them or, finally, the dependence may exist and remain unknown.

The general form of the theory admits of the existence of a dependence only in the sense that numbers, one related to the other, should be thought of as given together.

N. I. Lobačevskiĭ

Before entering upon stories about methods of investigation of problems of maxima and minima, we will discuss functions for a while.

The concept of a function is the key concept of mathematical analysis. But surprisingly, this concept was not formulated at once. At first it was vague and without a reasonably accurate description. The first attempts to outline the contours of the function concept were made at the end of the seventeenth century by Leibniz and Johann Bernoulli. Leibniz introduced the term "function." Bernoulli associated with this term the notion of *"an expression made up in some way out of a variable magnitude and constants."* Euler later made Bernoulli's idea more concrete, defining a function in his textbooks as *an analytic expression* made up of a variable magnitude and of constants. He also introduced the symbolism $f(x)$. Euler admitted the possibility of calling a function "any curve drawn freehand" and was aware of cases of functions—the sine function, for example—that can be described verbally.

Let's consider a simple example of a function that admits different descriptions. We have in mind the function $y = |x|$. $|x| = \sqrt{x^2} = (x^2)^{1/2}$ is undoubtedly an (analytic) expression made up of the variable magnitude x and of constants. As such it is a function as defined by Euler and by Johann Bernoulli. But it can also be represented as a "drawn curve" that follows the bisectors of the first and second quadrants in a rectangular coordinate system. And it can be described in words as the function that is equal to zero for x equal to zero, to x for positive x, and to $-x$ for negative x. Quite generally, most functions admit different descriptions.

Which should be the preferred description? This question gave rise to may disagreements. Euler thought that the class of functions that are "curves drawn freehand" is larger than the class of functions given by "analytic expressions." D'Alembert opposed this view and claimed that the two classes are the same.

Daniel Bernoulli advanced what seemed like a paradoxical view of the concept of function: namely, that an arbitrary *periodic* function with period 2π can be represented as a sum

$$\sum_{k=0}^{\infty}(a_k \cos kx + b_k \sin kx).$$

The majority of mathematicians felt that this is a rather restricted class of functions, more restricted than "analytic expressions," and, obviously, more restricted than arbitrary curves.

At the beginning of the nineteenth century there began to crystallize the idea of a function as a *correspondence*, a law under which the independent variable x is transformed into y, regardless of the nature of such a correspondence. One of the first scholars to support this view was Lobačevskiĭ. His thoughts in this connection serve as the epigraph for this story.

At about the same time similar views emerged in the French and German mathematical schools. Textbooks of mathematical analysis began to introduce definitions of function such as the following:

> Let x and y be the two given variables between whose values there exists a certain dependence. In general, one of the variables, say x, is regarded as the independent one. The value of x can be chosen arbitrarily, but for a given x the value of y is not arbitrary. Then we say that y is a function of x.
>
> *Vallée-Poussin*

A variable magnitude y is said to be a function of the variable magnitude x if to every value of x there corresponds a single definite value of y.

Nemyckiĭ, Sludskaya, Čerkasov

Such definitions cannot satisfy those who demand logical rigor (in general, the number of such people is not very great). After all, the term "function" in these cases is defined in terms of notions that are indeterminate and vague ("dependence," "law," "correspondence," and so on).

The creation of set theory brought with it a measure of calm. Its foundations were laid at the end of the nineteenth century by Georg Cantor. Now everything seemed to have found its proper place. In particular, one now defined a function as follows: let X and Y be two sets. The set F of pairs (x, y), $x \in X$, $y \in Y$, is called a function if for every $x \in X$ there is exactly one $y \in Y$ such that $(x, y) \in F$. Then we write $y = f(x)$.[*]

The concepts of set theory made a tremendous impression on many mathematicians who witnessed the birth of the new theory. Hilbert, one of history's greatest mathematicians, had this to say of set theory:

> I think that it is the highest manifestation of mathematical genius and one of the greatest achievements of man's purely spiritual activities.

Almost all contemporary mathematical works make use of the fundamental concepts and symbolism of set theory in one form or another.

As time passed, critical voices were heard. Contradictions emerged and heated arguments arose. At the beginning of the twentieth century all of the leading mathematicians (Poincaré, Hilbert, Hadamard, Weyl, Brouwer, and others) took part in the discussion of problems connected with the crisis in the foundations of mathematics. For some mathematicians, the set-theoretic definitions (in particular, the definition of a function) were unacceptably broad. These mathematicians were convinced that any functional dependence of interest from the practical point of view must necessarily be "constructive." In such cases there must be a distinct rule (or, to use a current expression, an algorithm) such that given x one can seek the required y. The debate gave rise to whole schools of mathematics that rejected set theory. A special "constructive" mathematics began to develop, a system at once similar to and dissimilar from the mathematics based on set theory with which mathematicians were already familiar.

Many scholars were greatly shaken by the crisis connected with the rejection of the set-theoretic conceptions by some mathematicians. In this

---

[*]Something close in meaning to this is contained in the last phrase of the quotation from Lobačevskiĭ that is one of the epigraphs for this story.

connection Hilbert commented that "No one will expel us from the paradise created for us by Cantor." Others heatedly debated this point of view.

Let us now turn to our problem: What is a function? Which is the preferred point of view? Where is the truth? A deep rethinking of these questions would take us too far afield. What must be said, however, is that all sufficiently definite descriptions of the function concept will do for general use, that is, for the solution of real-life problems, for physics, and, more generally, for applications of mathematics in the natural sciences, in technology, in economics, and so on. Throughout this book we will find that the kind of understanding of the function concept that began to take shape at the very sources of mathematical analysis will suffice for our basic purposes. Thus we will think of the function concept "in the manner of Bernoulli" as "an expression made up in some manner out of a variable magnitude and constants."

Functions differ in the number of their variables. We will first discuss functions of a single real variable. When referring to a function $y = f(x)$, we will always have in mind a specific rule for obtaining the number y from the number x. For example, "take a number, square it, add one to the square, and take the square root." This is the description of the function $y = \sqrt{1 + x^2}$, which, of course, is an "expression made up of a variable magnitude and constants." (Note that, to a large extent, Bernoulli's views agree with the views of modern "constructivists"; we need only replace the vague term "expression" by the term "algorithm," which can be given a very precise meaning. The meaning in question is, essentially, a precisely formulated prescription of what to do with the number x to obtain y.)

Functions of a single variable can be represented by means of graphs. To this end we orient the x-axis, as usual, horizontally form left to right and the y-axis vertically upward. We denote the intersection of the axes by the letter O. Then we choose a unit of length for the axes. Next, having chosen (a value for) x we lay it off on the x-axis and then lay off a segment of length $y = f(x)$ on the perpendicular through the point on the x-axis. The result is the graph of the function $y = f(x)$. In particular, the graph of the function $y = \sqrt{1 + x^2}$ is a hyperbola. (See Figure 9.1.)

Now let's define and represent some of the most important functions of one variable.

The simplest of all functions is the *constant* function

$$y \equiv c.$$

This function associates to each number x one and the same number c. For $c = 1$, we get the constant function $f(x) \equiv 1$ represented in Figure 9.2.

Next in complexity are the *linear* functions

$$y = bx.$$

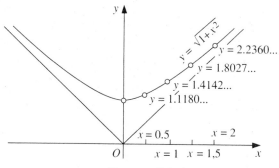

FIGURE 9.1

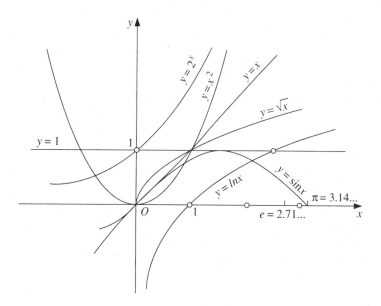

FIGURE 9.2

Here b is a constant and x is any real number. These functions are represented by lines through the origin other than the y-axis. For example, suppose $b = 2$. The corresponding linear function is $y = 2x$. Here, for $x = 1$, $y = 2$, for $x = 1/4$, $y = 1/2$, and so on; for each x the corresponding y is obtained in a definite manner, namely by multiplying x by 2. Figure 9.2 includes a representation of "the simplest" linear function $y = x$. Its graph is the bisector of the angle formed by the coordinate axes.

By combining constant and linear functions, we obtain functions of the form

$$y = bx + c.$$

In high school, these functions are called linear. We think that it is more natural to call them affine functions. Affine functions will play a very important role later in this book.

We consider next the *quadratic functions*

$$y = ax^2.$$

These functions are represented by parabolas passing through the origin. Take, for example, $a = 1/2$. Then for $x = 1$, $y = 1/2$, for $x = 2$, $y = 2$, for $x = 10$, $y = 50$, and so on; for each x the corresponding y is obtained in a definite manner, namely by multiplying x by itself and then multiplying the result by $1/2$. Figure 9.2 on page 85 shows the function

$$y = x^2 (a = 1).$$

By combining linear and quadratic functions, we obtain the *quadratic trinomials* $y = ax^2 + bx + c$.

Next come the *power functions*

$$y = Ax^n.$$

Here A is a real constant and n is a nonnegative integer.

The *exponential functions*

$$y = a^x,$$

play an important role in analysis. Figure 9.2 shows the function $y = 2^x$.

The functions that are inverses of the power and exponential functions are very useful. Figure 9.2 shows the functions $y = \sqrt{x}$ and $y = \ln x$. A singular feature of these functions is that they are not defined for all x. Thus $y = \sqrt{x}$ is defined only for nonnegative x, and $y = \ln x$ only for positive x (See Figure 9.2.)

The functions introduced in high school include the trigonometric functions. Figure 9.2 shows such a function, namely $y = \sin x$.

The following is a list of functions that we will constantly deal with hereafter:

The power function $y = Ax^n$; this is a constant function for $n = 0$, a linear function for $n = 1$, and a quadratic function for $n = 2$,

The "nth root of x" function $y = \sqrt[n]{x}$,

The exponential function $y = a^x$,

The trigonometric functions: the sine (function) $y = \sin x$, the cosine $y = \cos x$, the tangent $y = \sin x / \cos x = \tan x$, and the cotangent $y = \cos x / \sin x = \cot x$.

These functions can be combined to form various expressions, all of which are functions of one variable. Some relevant examples follow:

$$y = \sqrt{a^2 + x^2}, \qquad y = x\sqrt{1 - x^2}, \qquad y = \frac{ax}{2} - \frac{\pi x^3}{3},$$

$$y = \frac{(a + x)^2}{x}, \qquad y = \sqrt{b^2 + x^2 - 2bx \cos \alpha}.$$

Of course, this list can be extended. We have included only the functions that we will encounter when solving extremal problems. An example of a "scarier" function is

$$y = \sin\left(2^{\sqrt[5]{\log_7(1+\tan(x+1))}}\right).$$

Here I've encoded the following rule: take a number, add one to it, take the tangent of this sum, add one to the result, take the logarithm to the base 7 of latter sum, take the fifth root of the resulting number, raise two to the latter power, and take the sine of this power of two.

Similiar, but somewhat less scary expressions (or rules, or algorithms) exhaust the content of the notion of a "function of one variable," an idea that will be repeatedly encountered later in this story.

My handling of many important matters has been rather casual. I know that you have encountered functions of one variable in high school and I count on you to look in your textbook for more information, ask your teacher, or think some things through yourself.

All the same, functions of one variable won't suffice. We will find it indispensable to work with functions of two, three, four, a hundred variables (and—although you need not worry about it now—with functions of infinitely many variables). Never fear! None of this is terribly difficult.

At first we'll take just one step forward and discuss functions of *two variables*.

Many of you would be puzzled if asked when you first began to work with functions of two variables. Functions of two variables are not taught in school, so how could you have worked with them? But, in fact, we have all known about functions of two variables from time immemorial. Yes, immemorial! None of you can remember the day when, for the first time, your father, mother, or a visiting acquaintance asked you a question such as: "Here is one apple, and here is another—how many are there in all?" This was in your earliest childhood, when you could hardly talk and long before you learned the alphabet. You can't remember the time when you answered this question. But it was at that moment, when you added one apple to another apple and got "two" apples, that you first encountered a function of two variables, the oldest and best known of such functions, namely *addition*:

$$z = x + y.$$

The addition function associates to an arbitrary pair of numbers x and y their sum z. In particular, it associates to the pair $(1, 1)$ the number 2, to the pair $(7, -3)$ the number 4, and so on.

You learned first to add natural numbers, then integers, then real numbers. Thus, you associate to an arbitrary pair (x, y) of real numbers x and y their sum. Put differently, the addition operation is defined for all pairs

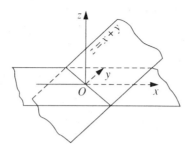

FIGURE 9.3

(x, y) of real numbers x and y. The point of all this is that you encountered functions of two variables far earlier than functions of one variable.

After addition, you learned subtraction, multiplication, and division. These operations are all functions of two variables. Subtraction and multiplication are defined for all pairs (x, y) of real numbers x and y; division is defined only for those pairs (x, y) for which $y \neq 0$.

Again following Bernoulli, by a function of two variables we will mean *an expression consisting of variable magnitudes x and y and of constants.*

Functions of two variables $z = f(x, y)$ can also be represented graphically. To this end we represent the (x, y)-plane and set the z-axis perpendicular to it. Now, given numbers x and y, we locate (x, y) in the (x, y)-plane, compute $f(x, y)$, and lay off a segment of length $f(x, y)$ on the line parallel to the z-axis and starting from the point (x, y).

Let's now define and represent some of the most important functions of two variables.

Again, the simplest function is the *constant* function

$$z \equiv c.$$

This function associates to each pair (x, y) of real numbers x and y the number c.

Next in complexity are the *linear* functions

$$z = ax + by.$$

Here a and b are constants and x and y are arbitrary real numbers. To find z given the pair (x, y) we must multiply x by a and y by b and add the resulting numbers. Suppose $a = 2$, and $b = 1/2$. Then to the pair $(1, 0)$ there corresponds the number $z = 2$, to the pair $(1, 1)$ the number $z = 5/2$, and to the pair $(4, -8)$ the number $z = 0$. Figure 9.3 shows the graph of the linear function $z = x + y$. The linear functions are represented by planes through the origin not perpendicular to the (x, y)-plane.

Using constant and linear functions, we form functions

$$z = ax + by + c.$$

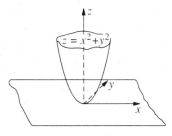

FIGURE 9.4

We will also call them *affine* functions. They will play an extremely important role later in the book.

Next come quadratic functions $z = Ax^2 + 2Bxy + Cy^2$ and functions of the form

$$z = Ax^2 + 2Bxy + Cy^2 + ax + by + c.$$

Figure 9.4 shows the function $z = x^2 + y^2$. Here are a few other examples of functions of two variables that we will encounter when solving various problems:

$$z = Ax^2(y - Bx), \qquad z = Axy,$$

$$z = \frac{x^2}{a^2} + \frac{y^2}{b^2}, \qquad z = (x - c)^2 + (y - d)^2,$$

$$z = \sqrt{(x - c)^2 + (y - d)^2}.$$

The fact that enables us to give a visual representation of the characteristic features of functions of two variables is that the graph of a remarkable function of two variables is always before our eyes and "under our feet."

Imagine that you are standing on the ground. Your position in space can be described by the triple of numbers (ϕ, θ, h), where φ and θ are, respectively, latitude and longitude, and h is *altitude above sea level*. Thus $h = h(\varphi, \theta)$—that is, h is a function of (φ, θ). (At this point, recall the general view of a function that we used as an epigraph for this story.) All that we see—hills, hollows, ravines, mountains, mirror-like surfaces of lakes—forms the "graph" of this function.

The graph of this function has many noteworthy points. Here, of course, the *peaks* come first. When we scale a peak we experience the joy of victory, the delight of surmounting a difficulty. When travelling in the mountains we look for *passes* that take us from valley to valley. When studying the sea bottom we try to find the deepest point in the sea. It is these very points—peaks, passes, and hollows—that will properly concern us later.

We will leave functions of two variables for now and take another step forward. We ask: What is a function of three variables? Of course, it is *an expression made up of three variables* x, y, z *and constants.* Examples? As many as you wish! The sum $u = x + y + z$, the product $u = x \cdot y \cdot z$, the constant function $u \equiv d$, the linear function $u = ax + by + cz$, the quadratic function $u = Ax^2 + 2Bxy + Cy^2 + 2Dyz + 2Gxz + Fz^2$, $u = \sqrt{x^2 + y^2 + z^2}$, $u = 2^x + \log_3 y + \sin z$, and so on, and so forth.

And what is a function of four variables? Obviously, *an expression made up of four variables* x, y, z, u *and constants.*

And what is a function of 26 variables? Of course, an expression made up of the variables, $a, b, c, d, e, f, g, h, i, j, k, l, m, n, o, p, q, r, s, t, u, v, w, x, y, z$ and constants.

And a function of 28 variables? Obviously, an expression made up of the variables...but, we've run out of letters! Are we to study only functions of at most 26 variables? If necessary we can throw in the Greek alphabet and some other alphabets to boot. But all this may not be enough. In modern economic problems there may be thousands of variables. What must we do?

The solution is simple. Instead of letters, the variables can be denoted by the single letter x with subscripts: $x_1, x_2, \ldots, x_n$. Thus a function of n variables can be defined as *an expression made up of* n *variables* $x_1, x_2, \ldots, x_n$, *and constants.*

The simplest function of n variables is a constant function $y \equiv c$. It associates to every choice of n numbers $(x_1, \ldots, x_n)$ the same number c.

One extremely important class is the class of *linear functions*

$$y = a_1 x_1 + \ldots + a_n x_n.$$

Constants and linear functions can be combined to form the *affine* functions

$$y = a_1 x_1 + \ldots + a_n x_n + c.$$

We will use functions of this kind to approximate more complicated functions.

This is perhaps the appropriate time to assign the proper names to certain functions. We have in mind functions of many variables. By now, you and I have encountered such functions many times. This happened, for example, in the fifth story in connection with means, where we considered functions such as

$$y = x_1 \ldots x_n, \qquad y = \sqrt[n]{x_1 \ldots x_n},$$
$$y = \sqrt{x_1^2 + \ldots + x_n^2}, \qquad (x_1^p + \ldots + x_n^p)^{1/p}, \qquad x_i \geq 0.$$

We also encountered functions of many variables in the story of the brachistochrone,

$$T_n = \frac{\sqrt{y_1^2 + x_1^2}}{\sqrt{2gy_1}} + \ldots +$$

$$\frac{\sqrt{(y_1 - y_2)^2 + (x_2 - x_1)^2}}{\sqrt{2gy_2}} + \ldots + \frac{\sqrt{(y_n - y_{n-1})^2 + (x_n - x_{n-1})^2}}{\sqrt{2gy_n}},$$

and in the story about Newton's problem. In the latter story we encountered even more striking functions, namely "functions of functions"—for example, functions of curves and thus, in effect, functions of infinitely many variables.

My aim in Part Two is to present a fragment of the general theory of extremal problems. As a first step, I have explained what functions of many variables are. I the next story I will explain how to pose maximum and minimum problems for functions of many variables subject to constraints.

10

What is an Extremal Problem?

In the first half of the book we solved a great many maximum and minimum problems. Some of them had been posed a long time ago, even centuries ago, and among their investigators were some of the greatest mathematicians of the past—Euclid, Archimedes, Kepler, Fermat, Bernoulli, Leibniz, and Newton. The investigations themselves took entirely dissimilar paths, and their aims were sometimes achieved only after long periods of meandering.

I have chosen as an epigraph for this part the words of d'Alembert. Ponder them. Is it possible to find a method for solving all problems (including those talked about in the first part) in one simple way?

Everybody realizes that there cannot be one simple rule for solving all problems in the world. (Incidentally, d'Alembert had in mind only problems in dynamics.) Even the possibility of the existence of a single method of solution of all problems discussed in the first half may strike some as doubtful. And yet such a method exists. We'll present it here, in the second half of the book. Using this one simple method, we'll solve all the problems from the first half in the same, standard, you might even say routine, way.

But first we must have a single way of writing down the conditions of problems and a general language for discussing problems of such different content. This is what the present story is about.

To begin, we'll once more make precise the meaning of the key words: "maximum," "minimum," "extremum," and "optimum."

Recall two planimetric problems from the first part. One was discussed on various occasions beginning with the first story, the other appeared in the fifth story.

HERON'S PROBLEM. *Given two points on the same side of a line, find a point on that line such that the sum of its distances to the given points is minimal. (Refer back to Figure 1.1.)*

KEPLER'S PLANIMETRIC PROBLEM. *Inscribe a rectangle of maximal area in a circle of unit radius.*

These problems differ in that in the first we must find a minimum and in the second, a maximum.

Recall that the words "maximum" and "minimum" are of Latin origin. They mean "largest" and "least," respectively. In connection with maximum and minimum problems we frequently use two other words of Latin origin. One of them is "extremum," meaning "extreme," a term that combines the concepts of maximum and minimum. (Its use was suggested by the French mathematician du Bois-Reymond.) The other is the adjective "optimal," derived from the Latin *optimus*, meaning "best" or "perfect." The term "optimal" has been universally adopted in recent years.

The theory of problems concerned with finding largest and least magnitudes is called *the theory of extremal problems* or *optimization theory*. If the problem involves finding the best influence on some processes and phenomena that man can control within some bounds, then we include it in the section of the theory of extremal problems called *optimal control*.

The Heron and Kepler problems were stated a few lines earlier using words rather than formulas. The same applies to all problems in the first part: the terminology used was geometric in geometric problems, (the isoperimetric, Steiner, and other problems), mechanical, in mechanical problems (the brachistochrone and Newton problems), and algebraic in algebraic problems (the Tartaglia problem, problems involving inequalities, and others). No formulas appeared in these problems. This was only proper. Extremal problems arising in mathematics, in the natural sciences, or in practical enterprises are traditionally stated first without formulas, using the terminology of the domain in which they arise. In order to be able to utilize a general theory, it is necessary to effect a *translation* of the statements of the problems from each specific language to the language of mathematics. Such a translation is called a *formalization*.

We will use examples to illustrate the process of formalization. Let's begin with Heron's problem.

We take the given line as the x-axis and draw the y-axis through the point A perpendicular to the x-axis. (Refer back to Figure 1.1.) Let the coordinates of the points A and B be $(0, a)$ and (d, b), respectively. On the x-axis we take a point D with coordinates $(x, 0)$. Then the sum of the distances from A to D and from D to B is $\sqrt{a^2 + x^2} + \sqrt{b^2 + (d - x)^2}$. This results in the following problem: *Find the least value of the function*

$$f(x) = \sqrt{a^2 + x^2} + \sqrt{b^2 + (d - x)^2}$$

for all values of x.

This formalization is very natural and almost forces itself upon us. But, in general, a problem can have many formalizations. We will illustrate this by means of the planimetric problem of Kepler.

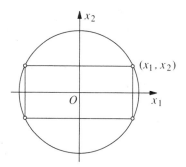

FIGURE 10.1

We orient the x_1 and x_2-axes parallel to the sides of the rectangle. The equation of the unit circle is $x_1^2 + x_2^2 = 1$. Let (x_1, x_2) be the coordinates of the vertex of the rectangle in the first quadrant. (See Figure 10.1.) Then the area of the rectangle is $4x_1 x_2$. This yields the following formalization: Find the largest value of the function of two variables

$$f_0(x_1, x_2) = 4x_1 x_2,$$

subject to the conditions

$$f_1(x_1, x_2) = x_1^2 + x_2^2 - 1 = 0, \; f_2(x_1, x_2) = x_1 \geq 0, \; f_3(x_1, x_2) = x_2 \geq 0.$$

Note that we can dispense with the inequalities $x_1 \geq 0$ and $x_2 \geq 0$. Then it is easy to see that the problem of finding the largest value of the function

$$f_0(x_1, x_2) = 4x_1 x_2,$$

subject to the condition

$$f_1(x_1, x_2) = x_1^2 + x_2^2 - 1 = 0,$$

is also a formalization of the planimetric problem of Kepler.

If we use the equation $f_1(x_1, x_2) = x_1^2 + x_2^2 - 1 = 0$ to express x_2 in terms of x_1 and substitute the result in the expression for f_0, then (after replacing x_1 by x) we obtain yet another formalization: *Find the largest value of the function $\varphi(x) = 4x\sqrt{1 - x^2}$ subject to the condition $0 \leq x \leq 1$ or the condition $|x| \leq 1$ (these conditions are dictated by the domain of definition of the function and by the sense of the variable x).

We see that there are different formalizations of the same problem. The ease of solution of a problem often depends on its clever formalization. Formalization is an art. It must be learned, and the best way to learn it is to solve practical problems.

We have talked about Kepler's planimetric problem. There are different ways of posing similar problems in space. We discussed one formulation in the first part—to inscribe a cylinder of maximal volume in a unit sphere. Another formulation is *to inscribe a rectangular parallelepiped of largest volume in a unit sphere*. This formulation yields the following formalization: Find the largest value of the function of three variables

$$f_0(x_1, x_2, x_3) = 8x_1 x_2 x_3,$$

subject to the condition

$$f_1(x_1, x_2, x_3) = x_1^2 + x_2^2 + x_3^2 - 1 = 0.$$

(Recall that Kepler considered the special case of this problem when $x_1 = x_2$).

The following problem leads to the same formalization (without the number 8): *Find the largest value of the product of three numbers subject to the condition that the sum of their squares is equal to a given number.* This problem can be generalized by replacing three with five, 10, or arbitrarily many numbers. The latter problem is formalized as follows: Find the largest value of the function of n variables

$$f_0(x_1, \ldots, x_n) = x_1 \cdots \cdots x_n,$$

subject to the condition

$$f_1(x_1, \ldots, x_n) = x_1^2 + \cdots + x_n^2 - 1 = 0.$$

We will now try to describe with a measure of definiteness all the elements of a correctly formalized extremal problem. It must, necessarily, involve a function (of n variables, say) for which we are to find the largest or least value (a function to be maximized or minimized), and a *constraint* given by a number of equalities and inequalities (with the same variables).

What follows is a list of some minimized or maximized functions and functions that define constraints, chosen from among functions already encountered or to be encountered.

$f_1(x) = \sqrt{a^2 + x^2} + \sqrt{b^2 + (d - x)^2}$ Heron's problem

$f_2(x) = \dfrac{\sqrt{a^2 + x^2}}{v_1} + \dfrac{\sqrt{b^2 + (d - x)^2}}{v_2}$ problem of reflection of light

$f_3(x) = \dfrac{H}{b} x(b - x)$ Euclid's problem

$g_1(x_1, x_2) = 4x_1 x_2,$
$g_2(x_1, x_2) = x_1^2 + x_2^2 - 1.$ Kepler's planimetric problem

$h_1(x_1, x_2, x_3) = 8x_1 x_2 x_3,$
$h_2(x_1, x_2, x_3) = x_1^2 + x_2^2 + x_3^2 - 1.$ the problem of a parallelepiped
$F_0(x_1, \ldots, x_n) = x_1 \cdots \cdots x_n,$ inscribed in a sphere
$F_1(x_1, \ldots, x_n) = x_1^2 + \cdots + x_n^2 - 1.$

Here f_1, f_2, and f_3 are functions of one variable, g_1 and g_2 of two variables, h_1 and h_2 of three variables, and F_0 and F_1 of n variables.

Thus to *formalize an extremal problem is to describe precisely a function* (denoted by f_0) *to be minimized or maximized and a constraint* (denoted by C.) The constraint is usually given by equalities and inequalities.

We will use the abbreviated notation*

(p) $$f_0(x) \to \min(\max) \quad \text{for } x \text{ in } C,$$

for the following formalized problem: "*Find the minimum* (*maximum*) *of the function* $f_0(x)$ *subject to the condition that* x *is in* C." The points in C are called *admissible*; if there are no constraints then (p) is called *a problem without constraints*. For example, the abbreviated description of one of the formalizations of Kepler's planimetric problem is

(p_1) $$f_0(x_1, x_2) = 4x_1 x_2 \to \max, \qquad f_1(x_1, x_2) = x_1^2 + x_2^2 - 1 = 0,$$

and of Heron's problem, simply,

(p_2) $$f_0(x) = \sqrt{a^2 + x^2} + \sqrt{b^2 + (d - x)^2} \to \min.$$

(p_1) *is a problem with a constraint involving an equality and* (p_2) *is a problem without constraints.*

An admissible point $\hat{x}$ is called *an absolute minimum* (*maximum*) of a problem (p) if $f(x) \geq f(\hat{x})$ for every x in C (if $f(x) \leq f(\hat{x})$ for every x in C). An absolute minimum (maximum) of a problem is called *a solution of the problem*. Our aim is to find a solution.

To find a solution we will resort to finding so-called *local extrema*.

When we get to the top of a hill or knoll in an otherwise flat locality, then we are at its highest point. But this doesn't mean that we have solved the problem of finding the highest point above sea level. The only people who have "solved" the latter problem are those who have scaled Mount Everest. That's the difference between absolute and local extrema.

We give a precise definition of the latter concept. We will say that a point $\hat{x} = (\hat{x}_1, \ldots, \hat{x}_n)$ yields a local minimum (maximum) for a problem (p) if there is a number $\varepsilon > 0$ such that for all points $x = (x_1, \ldots, x_n)$ in C for which

$$\sqrt{(x_1 - \hat{x}_2)^2 + \cdots + (x_n - \hat{x}_n)^2} < \varepsilon,$$

we have the inequality

$$f_0(x) \geq f_0(\hat{x}) \qquad (f_0(x_0) \leq f_0(\hat{x})).$$

---

*The letter p is the first letter in the word "problem." The letter x stands for the n-tuple $(x_1, \ldots, x_n)$. Hereafter we sometimes speak of "the point x." If $x = (x_1, \ldots, x_n)$ and a is a number, then $ax = (ax_1, \ldots, ax_n)$, and if $x' = (x_1', \ldots, x_n')$ then $x + x' = (x_1 + x_1', \ldots, x_n + x_n')$.

(In other words, if the value of the function at an admissible point "near" $\hat{x}$ does not exceed (is not less than) $f_0(\hat{x})$.)

We will end this story with a formalization of the transportation problem that we mentioned casually in the first story. Recall that in this problem we must set up a shipping schedule for sending a certain product from supply centers to stores at minimal transportation cost. Let a_i denote the number of units of freight at the ith supply center and m the number of centers. Let b_j, $j = 1, \ldots, n$, denote the number of units of freight required by the jth store. Finally, let c_{ij} denote the cost of transporting a unit of freight from the ith supply center to the jth store. The amount of freight transported from the ith base is $x_1 + \cdots + x_{in}$, and this number must not exceed a_i (no supply center can supply more than it has in stock). The amount of freight transported to the jth store is $x_{1j} + \cdots + x_{mj}$, and this number must be exactly equal to the requirement b_j of the store. This implies the following formalization of the transportation problem:

$$c_{11}x_{11} + \cdots + c_{1n}x_{1n} + \cdots + c_{m1}x_{m1} + \cdots + c_{mn}x_{mn} \to \min,$$
$$x_{i1} + \cdots + x_{in} \le a_i, \qquad i = 1, \ldots, m,$$
$$x_{ij} + \cdots + x_{nj} = b_j, \qquad j = 1, \ldots, n,$$
$$x_{ij} \ge 0, \qquad 1 \le i \le m, \qquad 1 \le j \le n.$$

We note that in the transportation problem the function to be minimized as well as the constraints are given by means of *linear functions*. The section of the theory of extremal problems where one studies extrema of linear functions subject to linear constraints is called *linear programming*. It includes the transportation problem.

The next story deals with extrema of functions of one variable. This topic is now covered in high school, but we won't be daunted by the prospect of a measure of repetition.

11

Extrema of Functions of One Variable

> When a quantity is greatest or least, at that moment
> its flow neither increases nor decreases.
>
> *I. Newton*

This story and the next one have the same two-part structure. In the first part of each story we present a solution method without proof but with some explanations and comments. In the second half, we give exact definitions and some proofs. To master the first part the reader needs to know only the concepts of "limit," "continuous function," and "derivative."

1. First, let's examine a method of solution of extremal problems for functions of one variable of the following type:

$$(p) \qquad f_0(x) \to \min(\max), \qquad a \le x \le b.$$

a and b in (p) can be infinite. This means that we will consider extrema of functions f_0 on a finite interval, on a ray, or on the totality of real numbers.

EXAMPLES.

$$(p_1) \qquad f_0(x) = \sqrt{a^2 + x^2} + \sqrt{b^2 + (d-x)^2} \to \min,$$

$$(p_2) \qquad f_0(x) = \sqrt{1 - x^2} \to \max, \qquad -1 \le x \le .$$

Recall that (p_1) is a formalization of Heron's problem and (p_2) is a formalization of Kepler's planimetric problem (both formalizations were given in the tenth story).

Not all problems are solvable. For example, we have already considered the following unconstrained problem:

$(\mathbf{p_3})$ $$f_0(x) = -\frac{1}{1+x^2} \to \max .$$

The function $f_0(x) \leq 0$ and there is no point $\overline{x}$ such that $f_0(\overline{x}) = 0$. On the other hand, if $x_n = n$, $n = 1, 2, \dots$, then $f_0(x_n) \to 0$. This means that $(\mathbf{p_3})$ has no maximum, that is, there is no point $\hat{x}$ such that $f_0(x) \leq f_0(\hat{x})$ for all x.

While maxima and minima need not always exist, the following theorem of Weierstrass guarantees the existence of solutions in a tremendous number of cases.

THE THEOREM OF WEIERSTRASS. *Let $f_0(x)$ be a continuous function on a finite interval $[a, b]$. Then there exist solutions of the problems*

$(\mathbf{p_{min}})$ $$f_0(x) \to \min, \qquad a \leq x \leq b,$$

and

$(\mathbf{p_{max}})$ $$f_0(x) \to \max, \qquad a \leq x \leq b.$$

An immediate consequence of this theorem is the existence of a solution of $(\mathbf{p_2})$. This cannot yet be claimed for $(\mathbf{p_1})$, where the function is considered on the whole line rather than on a finite interval.

The theorem of Weierstrass implies a corollary that will allow us, among other things, to prove the existence of a solution for $(\mathbf{p_1})$.

COROLLARY. *Let f_0 be continuous on the whole line. If $\lim_{x \to \infty} f_0(x) = \lim_{x \to -\infty} f_0(x) = \infty$. Then the unconstrained problem*

$$f_0(x) \to \min,$$

has a solution.

We will also encounter the case in which f_0 is continuous on a ray $a \leq x < \infty$ or $a < x < \infty$. If in the first case $\lim_{x \to \infty} f_0(x) = \infty$ and in the second case $\lim_{x \to a} f_0(x) = \lim_{x \to \infty} f_0(x) = \infty$, then the function f_0 attains its minimum on the corresponding ray.

To find the solution of the problem (p), we'll use a method first applied by Fermat. Before we solve the problem, however, we recall a definition introduced in the previous story. Let f_0 be a function defined on an interval $a \leq x \leq b$ and let $\hat{x}$ be a point in that interval. We say that $\hat{x}$ yields *a local minimum (maximum)* on (p) if there is an $\varepsilon > 0$ such that for all x in $[a, b]$ for which $|x - \hat{x}| < \varepsilon$, we have the inequality $f_0(x) \geq f_0(\hat{x})$ ($f_0(x) \leq f_0(\hat{x})$). We sometimes say more simply that $\hat{x}$ yields a local extremum of the function f_0.

We have the following theorem.

THEOREM OF FERMAT. *Let the function* f_0 *be differentiable at the point* $\hat{x}$. *If* $\hat{x}$ *yields a local extremum* (*minimum or maximum*) *of* f_0 *then* $f_0'(\hat{x}) = 0$.

Points such that $f_0'(x) = 0$ are called *stationary points*. Stationary points and endpoints are called *critical points*.

The relation $f_0'(\hat{x}) = 0$ is only a necessary condition for an extremum. Thus $\hat{x} = 0$ is a stationary point for the function $f_0(x) = x^3$ but yields neither a local maximum nor a local minimum.

Fermat's theorem gives rise to the following method for solving one-dimensional problems. We will divide the process into four stages.

FIRST STAGE. *This stage involves the formalization of the problem. If possible, the problem must be reduced to the form*

$$(p) \qquad\qquad f_0(x) \to \min(\max), \qquad a \le x \le b.$$

SECOND STAGE. *This stage includes setting down the necessary condition* $f_0'(x) = 0$.

THIRD STAGE. *This stage involves finding all stationary points.*

FOURTH STAGE. *This stage consists of sorting all critical values of* f_0 *and choosing the least* (*largest*) *of them.*

The Weierstrass and Fermat theorems imply that if a function f_0 satisfies the conditions of Weierstrass' theorem (or its corollaries) on $[a, b]$ and if the function is differentiable at the interior points x of the interval $[a, b]$ (for $a < x < b$), then the method outlined previously brings a solution of the problem.

One fact that we will use in our solution is that *if the segment* $[a, b]$ *is finite, and the function* f_0 *is continuous on* $[a, b]$ *and differentiable at its interior points* x, $a < x < b$, *then the solution is found among the critical points* (that is, at a point where the derivative is zero or at an endpoint).

This shows that in order to apply the rule just given we must know how to differentiate. To facilitate this procedure we adduce a table (see Table 11.1 on page 102) of derivatives of basic functions.

In addition to Table 11.1, it helps to remember the following formulas:

$$(7) \quad (f + g)' = f' + g', \qquad (fg)' = f'g + fg', \qquad \left(\frac{f}{g}\right)' = \frac{f'g - g'f}{g^2}.$$

In addition, we will frequently encounter functions of the form $h(x) = f(g(x))$. You should remember and learn to use the following formula ("the chain rule") for the derivative of such composite functions:

$$(8) \qquad\qquad h'(x) = f'(g(x))g'(x).$$

EXAMPLE. $h(x) = \sqrt{a^2 + x^2}$. Here $h(x) = f(g(x))$, where $f(u) = \sqrt{u} = u^{1/2}$, $g(x) = a^2 + x^2$. Using formula (1) from the table and formulas (7)

TABLE 11.1. DERIVATIVES

	$f(x)$	$f'(x)$
(1)	$x^a (a \neq 0, x > 0)$	ax^{a-1}
(2)	$a^x (a \neq 1 ; a > 0)$	$a^x \ln a$
(3)	$\log_a x (a \neq 1)$	$\frac{1}{x \ln a}$
(4)	$\ln x$	$\frac{1}{x}$
(5)	$\sin x$	$\cos x$
(6)	$\cos x$	$- \sin x$

and (8) we obtain

$$h'(x) = \frac{1}{2\sqrt{a^2 + x^2}} \cdot (a^2 + x^2)' = \frac{2x}{2\sqrt{a^2 + x^2}} = \frac{x}{\sqrt{a^2 + x^2}}.$$

We will conclude this section with a few words about *convex functions*. These functions play a very important role in the theory of extremal problems, and we'll find it necessary to recall this topic many times. Now let's give a definition of a convex function of one variable.

You may have encountered the notion of convexity already in high school. Recall that a figure is called *convex* if it contains the interval between any two of its points. Viewed as part of the plane, any triangle is a convex figure, but there are nonconvex quadrilaterals. (See Figure 11.1.)

One can give three equivalent definitions of a convex function. One definition is that a function $y = f(x)$ is *convex* if for any chord joining two points on its graph, the part of the graph corresponding to the intermediate points lies below the chord. A second definition is that f is convex if the set of points above the graph is convex. The third definition of convexity of f requires that for any numbers x_1 and x_2 and any $\alpha, 0 \leq \alpha \leq 1$, the

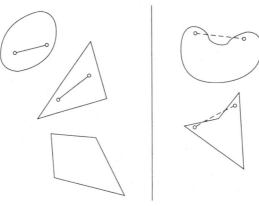

FIGURE 11.1

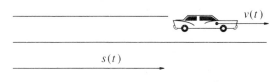

$$\text{FIGURE } 11.2$$

following inequality (*Jensen's inequality*) holds:

$$f(\alpha x_1 + (1 - \alpha)x_2) \le \alpha f(x_1) + (1 - \alpha)f(x_2).$$

All linear functions, functions of the form $y = bx + c$ (affine functions), and quadratic trinomials $y = ax^2 + bx + c$ with $a > 0$ are convex. Of the functions $y = |x|^p$ only those with $p \ge 1$ are convex. The function $y = \sqrt{h^2 + y^2}$ is convex for all h. Not all convex functions are differentiable everywhere. For example, the function $y = |x|$ is not differentiable at zero. But if a convex function is differentiable, then its derivative is an increasing function.

2. In the first half of this story we have described a rule for the solution of problems. This rule is easy to remember and, using it, we can immediately solve problems (which is what we will be doing in the thirteenth story). We are sure, however, that many readers will want to know about the origin of this rule. We will explain this not once but twice. You might easily ask why.

I think that readers who enjoy scientific and popular-scientific literature fall into two categories. One category consists of the majority of readers who try to understand just the fundamental ideas. They like presentations that are expressive, if not quite rigorous, and don't complain if they notice that apparently inessential details are missing. It is this category of readers that I had in mind in the first half of this story and will have in mind in the first half of the next story.

But a writer must not forget those readers who are not satisfied with the description of general ideas alone, readers who insist on getting to the essence of ideas, to the very heart of the matter, if possible. The concluding part of this section has been written with this category of readers in mind. In this part we will try to be as precise and brief as possible.

Imagine driving along a rectilinear highway. (See Figure 11.2.) At every moment your car is at some definite distance from some initial point. This means that the location of the car can be given at every moment t by a single number $s(t)$. In this way we obtain *a function of time*: $s(t)$ is the *distance* at time t from the car to the initial point.

Now look at the speedometer. It indicates *velocity*. We denote the velocity of the car at time t by $v(t)$. From courses in physics and mathematics we know that *velocity is the* time *derivative* of the distance function $s(t)$:

$$v(t) = \frac{ds}{dt} = s'(t).$$

(Some say that Lord Kelvin, one of the best physicists of the nineteenth century, claimed the opposite. He would say something like "Don't bother me with your mathematics: *the derivative is velocity!*")

If at a certain moment the velocity is not zero—assume, for definiteness, that it is positive, as shown in Figure 11.2 on p. 103—then we will be further away from the initial point during the subsequent moments, just as we were closer to it in the preceding moments. This means that at the moment in question the distance function $s(t)$ can have neither a maximum nor a minimum. It follows that *at a maximum or minimum point the velocity must be zero.* But this is just what Fermat's theorem is about.

Now we say the same thing in a more rigorous way. Let's begin with a precise definition of the derivative. We could use the definition presented in high school, but this would lead to difficulties in subsequent stories, when we talk of derivatives of functions of many variables. Thus, we will give a definition that is equally applicable in the finite-dimensional case and in the infinite-dimensional case (which we'll take up later).

What does it mean to say that *a function f is differentiable at a given point x_0* (or, equivalently, *has a derivative at x_0*)? If we avoid formulas, then we can say that this means that the function $f(x_0 + x) - f(x_0)$ *is closely approximated by a linear function.* If we are to be precise then this means the following.

DEFINITION. A function $y = f(x)$, defined on an interval $[a, b]$ that contains in its interior a point $x_0 (a < x_0 < b)$, is said to be *differentiable at x_0* (or, equivalently, *to have a derivative at x_0*) if there exists a linear function $y = kx$ such that

$$f(x_0 + x) - f(x_0) = kx + r(x),$$

where $\lim_{x \to 0} |r(x)|/|x| = 0$ (or, as is sometimes said, $r(x)/x$ is an infinitesimal).

An immediate consequence of our definition is that

$$k = \lim_{x \to 0} \frac{f(x_0 + x) - f(x_0)}{x},$$

which means that the number k in the definition is uniquely determined. This number is called the derivative of f at the point x_0 and is denoted by $f'(x_0)$.

The geometric sense of the derivative is that the line that is the graph of the function $y = f'(x_0)(x - x_0) + f(x_0)$ (this line passes through the point $(x_0, f(x_0))$ and its slope is equal to the derivative $f'(x_0)$) is tangent to the graph of the function $y = f(x)$. (See Figure 11.3.)

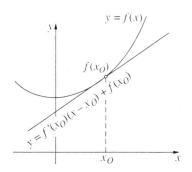

FIGURE 11.3

EXAMPLE 1. The quadratic trinomial $y = ax^2 + bx + c$ is differentiable everywhere and its derivative at x_0 is equal to $2ax_0 + b$. We will check this in the case of the function $f(x) = x^2$. We have

$$f(x_0 + x) - f(x_0) = (x_0 + x)^2 - x_0^2 = 2x_0 x + x^2.$$

Here $2x_0 x$ is a linear function and $r(x) = x^2$. Since $\lim_{x \to 0} |r(x)|/|x|$ $= 0$, f is differentiable at x_0 and $f'(x_0) = 2x_0$.

From the list given earlier in this story, we know that the elementary functions $a^x (a \neq 0)$, $\sin x$, $\cos x$, $\log_a x$ are differentiable wherever they are defined.

Let's look at an example of a function that is not differentiable at some point.

EXAMPLE 2. The function $y = |x|$ is not differentiable at zero. In fact, take any linear function $y = kx$. Assume for definiteness that $k \leq 0$. Put

$$r(x) = f(x) - f(0) - kx = |x| - kx = \begin{cases} x + |k|x, & x \geq 0 \\ -x + |k|x, & x \leq 0. \end{cases}$$

This means that $\lim_{x \to 0, x \geq 0} |r(x)|/|x| = 1 + |k| \neq 0$; thus, the function is not differentiable. The case $k \geq 0$ is similar.

FERMAT'S THEOREM. *Let $f_0(x)$ be a function defined on an interval $[a, b]$ that contains in its interior a point $\hat{x} (a < \hat{x} < b)$ and differentiable at $\hat{x}$. If $\hat{x}$ yields a local extremum (minimum or maximum) of this function then $f'(\hat{x}) = 0$.*

A precise definition of a local extremum was given earlier in this story.

PROOF. We assume that $f_0'(\hat{x}) = k \neq 0$ and show that $\hat{x}$ is not a local extremum. We suppose that $k > 0$. By the definition of limit, the fact that $\lim_{x \to 0} |r(x)|/|x| = 0$ (where $r(x) = f_0(\hat{x} + x) - f_0(\hat{x}) - kx$) implies that there is $\delta > 0$ such that if $|x| < \delta$ then $|r(x)| < (k/2)|x|$. But then for $x > 0$, $r(x) \geq -(k/2)x$, so that

$$f_0(\hat{x} + x) = f_0(\hat{x}) + kx + r(x) \geq f_0(\hat{x}) + kx - \frac{k}{2}x = f_0(\hat{x}) + \frac{k}{2}x > f_0(\hat{x}),$$

and for $x < 0$, $r(x) \le -(k/2)x$, so that

$$f_0(\hat{x} + x) = f_0(\hat{x}) + kx + r(x) \le f_0(\hat{x}) + kx - \frac{k}{2}x = f_0(\hat{x}) + \frac{kx}{2} < f_0(\hat{x}).$$

In other words, to the left of $\hat{x}$ the value of f_0 is less than $f_0(\hat{x})$ and to the right of $\hat{x}$ it is greater than $f_0(\hat{x})$. This means that $\hat{x}$ is neither a maximum nor a minimum. This completes the proof.

The geometric sense of Fermat's theorem is that *at a maximum or minimum point the tangent is horizontal.* We also want to emphasize the "computational" sense of an extremum that Kepler talks about (see the epigraph to the sixth story). Consider, for example, the functions $f_1(x) = x$ and $f_2(x) = x^2$. The first does not have an extremum at zero; the second does. If we increase the argument then the first function changes by the same amount and the second undergoes "imperceptible changes." Specifically, if $x = .01$ (which can still be represented on millimeter paper), then $f_2(x) = .0001$, and this is altogether "imperceptible."

This is how things are with Fermat's theorem. We postpone the exposition of some historical material to the fourteenth story.

It remains to prove Weierstrass' theorem on the existence of an extremum of a continuous function on a bounded interval. Since any interval can be transformed into the unit interval $[0, 1]$, it is in the latter interval with which we will work from now on.

To begin, we will prove the following lemma on a monotonic sequences of numbers.

LEMMA. *Every monotonic sequence of numbers in the unit interval has a limit in this interval.*

This means that if an infinite sequence of numbers $\{x_1, \ldots, x_n, \ldots\}$ is such that all of its elements belong to the unit interval (that is $0 \le x_n \le 1$, $n = 1, 2, \ldots$), and, furthermore, this sequence is monotonic (say, monotonically increasing, that is, $x_1 \le x_2 \le \cdots \le x_n \le \cdots$), then there is a number x_0 in the unit interval ($0 \le x_0 \le 1$) such that $\lim_{n\to\infty} x_n = x_0$.

Before proving the lemma, we point out that a number in the unit interval is representable by an infinite decimal $0.n_1 n_2 n_3 \cdots$ where n_i is one of the ten digits $0, 1, 2, 3, 4, 5, 6, 7, 8, 9$.

PROOF. Consider the first digit after the decimal point in each of the decimal representations of the numbers in our sequence $\{x_1, \ldots, x_n, \ldots\}$. These are integers not less than 0 and not greater than 9. Since our sequence is monotonically increasing, these integers likewise form a monotonically increasing sequence. One of these integers, denote it by n_1, *must repeat infinitely many times.* Let x_{N_1} be the first number in $\{x_1, \ldots, x_n, \ldots\}$ whose first digit after the decimal point is n_1. Then our sequence of integers cannot contain an integer greater than n_1; otherwise, in view of the monotonicity of our sequence, n_1 could not reappear.

Next we consider the second digit after the decimal in each of the numbers $\{x_{N_1}, x_{N_1+1}, \dots\}$. These again form a monotonically increasing sequence of integers not less than 0 and not greater than 9. We again take the integer n_2 that appears infinitely many times and the first number x_{N_2} of the new sequence whose second digit after the decimal sign is n_2. By continuing in this way, we obtain an infinite decimal $0.n_1 n_2 \dots$ that represents some number x_0 in the unit interval. Beginning with x_{N_1}, all numbers in the original sequence are of the form $.n_1 \dots$. Beginning with x_{N_2} all numbers are of the form $.n_1 n_2 \dots$, and so on. It follows that for $n = 1, 2, \dots$

$$x_n \leq x_0 \Leftrightarrow x_n - x_0 \leq 0,$$

and that for $n > N_s$,

$$x_0 - x_n \leq 10^{-s}.$$

This implies that $\lim_{n \to \infty} x_n = x_0$, which is what we wished to prove.

WEIERSTRASS' THEOREM. *A continuous function on a finite interval takes on its maximal and minimal values.*

We recall that a function $y = f(x)$ defined on an interval $[a, b]$ containing a point x_0 $(a \leq x_0 \leq b)$ is said to be *continuous at the point* x_0 if for every $\varepsilon > 0$ there is a $\delta > 0$ such that $|x - x_0| < \delta$, $a \leq x \leq b$, implies that $|f(x) - f(x_0)| < \varepsilon$. An immediate consequence of this definition is that if f is continuous at x_0 and $\{x_1, \dots, x_n, \dots\}$ is a sequence converging to x_0 ($\lim_{n \to \infty} x_n = x_0$), then the sequence $\{f(x_1), \dots, f(x_n), \dots\}$ converges to $f(x_0)$ ($\lim_{n \to \infty} f(x_n) = f(x_0)$). A function is said to be *continuous on an interval* if it is continuous at each point of this interval. A function $y = f(x)$ defined on $[a, b]$ is said to take on at the point x_0 its maximal (minimal) value on $[a, b]$ if $f(x_0) \geq f(x)$ $(f(x_0) \leq f(x))$ for all x in $[a, b]$.

We can now prove Weierstrass' theorem. We will prove it for a maximum.

PROOF. Let the function $y = f(x)$ be defined and continuous on the unit interval $[0, 1]$. Take two intervals $\Delta_1 = [a_1, b_1]$ and $\Delta_2 = [a_2, b_2]$ in $[0, 1]$. We will say that Δ_1 is *better* than Δ_2 if there is a point $\overline{x}$ in Δ_1 such that $f(\overline{x}) > f(x)$ for all x in Δ_2.

We divide the interval $\Delta^\circ = [0, 1]$ into two equal intervals $\Delta_1^1 = [0, 1/2]$ and $\Delta_2^1 = [1/2, 1]$.

We choose the better of the intervals Δ_1^1 and Δ_2^1; if neither is better, then we choose either one of the two. We denote by x_1 the left endpoint of the selected interval Δ^1.

We claim that, in view of our choice, for each point x not in Δ^1 there is a point $\overline{x}$ (that may depend on x) in Δ^1 such that $f(\overline{x}) \geq f(x)$. In fact, if Δ^1 is better, then our proof is complete. If Δ^1 is not better and there is no such $\overline{x}$, then this implies that the other interval is better, and this contradicts our choice.

Next, we divide the interval Δ^1 into two equal intervals Δ_1^2 and Δ_2^2 and again choose the better interval or either. We denote by x_2 the left endpoint of the selected interval Δ^2. In view of our choice, we can again claim that for each point x not in Δ^2 there is a point $\overline{x}$ in Δ^2 such that $f(\overline{x}) \geq f(x)$ (think this through).

Proceeding in this manner, we end up with a monotonic sequence $\{x_1, \ldots, x_n, \ldots\}$ of elements in $[0, 1]$. In view of our lemma, this sequence converges to some limit x_0. We'll prove that $f(x_0) \geq f(x)$ for all x in $[0, 1]$. In fact, assume that for $\tilde{x}$, $f(\tilde{x}) > f(x_0)$. Choose δ so small that $|x_0 - \tilde{x}| > \delta$ and that $|x - x_0| < \delta$, $0 \leq x \leq 1$, implies $f(x) < f(\tilde{x})$. The lengths of the intervals Δ^n are 2^{-n} and their left endpoints tend to x_0. This means that, at some moment, the whole interval Δ^n is in the interval $(x_0 - \delta, x_0 + \delta)$. But then, on the one hand, this Δ^n contains a point $\overline{x}$ such that $f(\overline{x}) \geq f(\tilde{x})$ and, on the other hand (since $|\overline{x} - x_0| < \delta$), $f(\overline{x}) < f(\tilde{x})$. This contradiction proves our theorem.

It is now easy to prove the corollary of our theorem (formulated in the early part of this story).

Let A be a number such that for $|x| \geq A$ we have $f(x) \geq f(0)$. By Weierstrass' theorem, there is a point x_0 in $[-A, A]$ such that $f(x_0) \leq f(x)$ for all x in $[-A, A]$. In particular, $f(x_0) \leq f(0)$. For $|x| > A$, $f(x) \geq f(0) \geq f(x_0)$. Hence $f(x) \geq f(x_0)$ for all x, which was to be proved. The two remaining cases (involving rays) are just as easy to prove.

Thus all the facts taken up in the first section of this story have been established.

12

Extrema of Functions of Many Variables. The Lagrange Principle

One can state the following general principle. If one is looking for the maximum or minimum of some function of many variables subject to the condition that these variables are related by a constraint given by one or more equations, then one should add to the function whose extremum is sought the functions that yield the constraint equations each multiplied by undetermined multipliers and seek the maximum or minimum of the resulting sum as if the variables were independent. The resulting equations, combined with the constraint equations, will serve to determine all unknowns.

J. Lagrange

1. In this story we will discuss ways to solve extremum problems for functions of many variables. The essence of the matter is expressed by Lagrange in the words we chose as an epigraph. To understand the first part of this story one must be familiar with the concept of "a continuous function of many variables" (discussed in the tenth story) and the material in the first part of the eleventh story.

Let $f_0, f_1, \ldots, f_m$ be functions of n variables $x = (x_1, \ldots, x_n)$. In principle, we will consider problems where the constraints are equalities as well as inequalities:

(p) $\qquad$ $f_0(x) \to \min(\max)$, $\qquad$ $f_i(x) = 0$, $\qquad$ $i = 1, \ldots, m'$,

$\qquad\qquad\qquad\qquad\qquad\quad f_i(x) \le 0$, $\qquad$ $i = m' + 1, \ldots, m$.

For the most part, however, we will deal with problems where the constraints are equalities only. Examples are

$$f_0(x) = (x_1 - a_1)^2 + (x_2 - a_2)^2 + (x_1 - b_1)^2 + (x_2 - b_2)^2 + (x_1 - c_1)^2$$

(p_1) $\qquad$ $+(x_2 - c_2)^2 \to \min$, $\qquad$ $x = (x_1, x_2)$.

(p_2) $\qquad$ $f_0(x) = x_1 \cdots x_n \to \max$, $\qquad$ $f_1(x) = x_1^2 + \cdots + x_n^2 - 1 = 0$.

Recall that for $n = 2$, (p_2) is the planimetric Kepler problem and for $n = 3$ it is the classical Kepler problem of the parallelepiped of maximal volume inscribed in a sphere. As for (p_1), it is the formalization of the following problem: *Find the point in the plane such that the sum of the squares of its distances from three given points is a minimum* (compare this with the Steiner problem).

Obviously, not every problem of type (p) has a solution. However, just as in the case of a function of one variable, it is possible to formulate a theorem (also proved by Weierstrass) that guarantees the existence of a solution in many cases.

Let C denote the set of admissible points in problem (p). This means that C consists of the points x such that

$$f_i(x) = 0, \qquad i = 1, \ldots, m', \qquad f_i(x) \le 0, \qquad i = m' + 1, \ldots, m.$$

The set C is said to be bounded if there is a constant $A > 0$ such that $|x_i| \le A$, $i = 1, \ldots, n$, for all $x = (x_1, \ldots, x_n)$ in C. For example, the set $x_1^2 + \cdots + x_n^2 = 1$ is bounded (in particular, the circle $x_1^2 + x_2^2 = 1$ is bounded) and the set $x_1 = x_2^2$ (a parabola) is unbounded.

THE THEOREM OF WEIERSTRASS. *Assume that the functions $f_0, \ldots, f_m$ in problem (p) are continuous and the set C of admissible points in (p) is bounded. Then the problems*

($p_{\min}$) $\qquad$ $f_0(x) \to \min$, $\qquad$ $f_i(x) = 0$, $\qquad$ $i = 1, \ldots, m'$,

$\qquad\qquad\qquad\qquad\qquad\quad f_i(x) \le 0$, $\qquad$ $i = m' + 1, \ldots, m$,

and

($p_{\max}$) $\qquad$ $f_0(x) \to \min$, $\qquad$ $f_i(x) = 0$, $\qquad$ $i = 1, \ldots, m'$,

$\qquad\qquad\qquad\qquad\qquad\quad f_i(x) \le 0$, $\qquad$ $i = m' + 1, \ldots, m$

are solvable.

As in the previous story, we note the following corollary to this theorem.

COROLLARY. *If the function* $f_0(x)$ $(x = (x_1, \ldots, x_n))$ *is continuous for all* x *and* $\lim f_0(x) = \infty$ *for* $x_1^2 + \cdots + x_n^2 \to \infty$, *then the unconstrained problem*:

$$f_0(x) \to \min,$$

is solvable.

We repeat the definition of a local extremum in problem (p).

DEFINITION. A point $\hat{x} = (\hat{x}_1, \ldots, \hat{x}_n)$ is said to yield *a local minimum* (*maximum*) in problem (p) if there is an $\varepsilon > 0$ such that for all admissible points $x = (x_1, \ldots, x_n)$ for which

$$|x_i - \hat{x}_i| < \varepsilon, \qquad i = 1, \ldots, n,$$

the inequality $f_0(x) \geq f_0(\hat{x})$ $(f_0(x) \leq f_0(\hat{x}_0))$ holds.

If $\hat{x}$ yields a local extremum for the unconstrained problem (p) then we also say that $\hat{x}$ yields a local extremum of the function f_0.

Before we can formulate the fundamental rule for the solution of problems of type (p), we must introduce one more concept.

Let $y = f(x)$ be a function defined for all $x = (x_1, \ldots, x_n)$ satisfying the inequalities $a_j \leq x_j \leq b_j$, $j = 1, \ldots, n$ (the set of such points is called the parallelepiped Π $(a_1, b_1; \ldots; a_n, b_n))$, and let $x_0 = (x_{01}, \ldots, x_{0n})$ be a point satisfying the strict inequalities $a_j < x_j < b_j$, $j = 1, \ldots, n$. We will consider the following function of one variable:

$$g_j(x) = f(x_{01}, \ldots, x_{0,j-1}, x_{0,j} + x, x_{0,j+1}, \ldots, x_{0n}).$$

What have we done? We have fixed all but the jth coordinate of x_0 and added x to the jth coordinate. Now we assume that the function g_j is differentiable at zero.

DEFINITION 2. The derivative at zero of the function g_j is called the jth *partial derivative of the function* f *at the point* x_0 and is denoted by $\partial f(x_0)/\partial x_j$.

Now we formulate two theorems that will enable us to state a rule for the solution of the problems of type (p). We'll consider first the unconstrained problem (p).

FERMAT'S THEOREM. *Suppose that all partial derivatives of the function* f_0 *exist at the point* $\hat{x}$. *If* $\hat{x}$ *yields a local extremum* (*minimum or maximum*) *of* f_0 *then*

(1) $$\frac{\partial f_0(\hat{x})}{\partial x_j} = 0, \qquad j = 1, \ldots, n.$$

Points at which all partial derivatives vanish are called *stationary*. Of course, just as in the case $n = 1$, condition (1) is a *necessary condition*.

By way of illustrating the use of this theorem, we will solve the problem (p_1). The corollary of Weierstrass' theorem guarantees the existence of a solution, which we denote by $(\hat{x}, \hat{y})$. We have

$$0 = \frac{\partial f_0(\hat{x}, \hat{y})}{\partial x_1} = 2[(\hat{x} - a_1) + (\hat{x} - b_1) + (\hat{x} - c_1)] \Rightarrow \hat{x} = (a_1 + b_1 + c_1)/3,$$

$$0 = \frac{\partial f_0(\hat{x}, \hat{y})}{\partial x_2} = 2[(\hat{y} - a_2) + (\hat{y} - b_2) + (\hat{y} - c_2)] \Rightarrow \hat{y} = (a_2 + b_2 + c_2)/3.$$

Hence the answer: $(\hat{x}, \hat{y})$ is *the center of gravity of the triangle* (a_1, a_2), (b_1, b_2), (c_1, c_2).

Now we will consider the general problem (p), except that *we will leave out inequalities*. We form the sum

$$\mathscr{L}(x, \lambda) = \lambda_0 f_0(x) + \lambda_1 f_1(x) + \cdots + \lambda_m f_m(x),$$

where $x = (x_1, \ldots, x_n)$ and $\lambda = (\lambda_0, \lambda_1, \ldots, \lambda_m)$. We call $\mathscr{L}(x, \lambda)$ the *Lagrange function* and the numbers $\lambda_0, \lambda_1, \ldots, \lambda_m$ the *Lagrange multipliers*.

The following is an abbreviated version of the general Lagrange principle in our epigraph: *in order to solve problem* (p) *(with equalities only) one forms the Lagrange function and treats it as if the variables* $x_1, \ldots, x_n$ *were independent* (that is, one applies Fermat's theorem). Then one solves the resulting equations

$$(2) \qquad \frac{\partial \mathscr{L}(x, \lambda)}{\partial x_j} = 0, \qquad j = 1, \ldots, n,$$

supplemented by the constraint equations

$$(3) \qquad f_i(x) = 0, \qquad i = 1, \ldots, m,$$

with respect to the variables $x_1, \ldots, x_n, \lambda_0, \lambda_1, \ldots, \lambda_m$ and selects from among these solutions the required one.

This rule is based on the following theorem.

THEOREM (the Lagrange multiplier rule). *Let* $f_0, \ldots, f_m$ *be functions defined in a parallelepiped* $\Pi(a_1, b_1; a_2, b_2; \ldots; a_n, b_n)$ *that contains in its interior a point* $\hat{x} = (\hat{x}_1, \ldots, \hat{x}_n)$ $(a_i < \hat{x}_i < b_i, i = 1, \ldots, n)$. *Also, let all functions* f_i, $i = 0, 1, \ldots, m$, *and all partial derivatives* $\partial f_i / \partial x_j$, $i = 0, 1, \ldots, m$; $j = 1, \ldots, n$, *be continuous in this parallelepiped. If the admissible point* $\hat{x}$ *yields a local extremum (minimum or maximum) then there are numbers* $\lambda_0, \lambda_1, \ldots, \lambda_m$, *not all zero, such that*

$$\frac{\partial \mathscr{L}(\hat{x}, \lambda)}{\partial x_j} = 0, \qquad j = 1, \ldots, n,$$

$$(\hat{x} = (\hat{x}_1, \ldots, \hat{x}_n), \quad \lambda = (\lambda_0, \lambda_1, \ldots, \lambda_m)).$$

Two observations are in order. First, while the system (2)–(3) contains $n + m$ equations in $n + m + 1$ unknowns, it must be borne in mind that *the Lagrange multipliers can be multiplied by any nonzero constant*. Since we can always multiply so that one of the Lagrange multipliers is 1, we can say that in

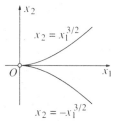

FIGURE 12.1

the system (2)–(3) *the number of equations is actually the same as the number of unknowns*. A second observation is that equations (2) are most meaningful if $\lambda_0 \neq 0$. Indeed, if $\lambda_0 = 0$, then equations (2) simply reflect the degeneracy of the constraints and are not related to the function whose extremum is sought. Usually one imposes additional constraints to ensure that $\lambda_0 \neq 0$ (in the theorem just stated, for $m = 2$, a sufficient condition is that the vectors $(\partial f_1(\hat{x})/\partial x_1, \ldots, \partial f_1(\hat{x})/\partial x_n)$ and $(\partial f_2(\hat{x})/(\partial x_1), \ldots, \partial f_2(\hat{x})/(\partial x_n))$ are not proportional). But one must not assume a priori that $\lambda_0 \neq 0$ (this is what Lagrange does—see the epigraph). The following example shows that the Lagrange multiplier rule may fail if we make the additional assumption that $\lambda_0 \neq 0$.

EXAMPLE. We consider the problem (Figure 12.1):

$$x_1 \to \min, \qquad x_2^2 - x_1^3 = 0.$$

Figure 12.1 shows that the only solution of the problem is the point $\hat{x} = (0, 0)$. We try to form the Lagrange function with $\lambda_0 = 1$ and apply the Lagrange algorithm:

$$\mathscr{L} = x_1 + \lambda(x_2^2 - x_1^3),$$

and

$$\frac{\partial \mathscr{L}}{\partial x_1} = 0 \Rightarrow -3\lambda x_1^2 + 1 = 0, \qquad \frac{\partial \mathscr{L}}{\partial x_2} = 0 \Rightarrow 2\lambda x_2 = 0.$$

The constraint equation is $x_2^2 - x_1^3 = 0$. The resulting system is obviously inconsistent.

The Lagrange multiplier rule yields the following recipe for looking for solutions of problem (p) with equalities. We will divide it into four stages.

The first stage is the formalization of the problem. Here we try (if possible) to reduce the problem to the form (p) with $m' = m$. The second stage is the application of the Lagrange principle, that is, setting down the system of equations

$$\frac{\partial \mathscr{L}}{\partial x_j} = 0, \qquad j = 1, \ldots, n,$$
$$f_i(x) = 0, \qquad i = 1, \ldots, m.$$

The third stage is finding all stationary points. Here it may be useful to first clarify the question of whether λ_0 can equal zero or not.

The fourth stage is selecting from among the stationary points the points where the function f_0 takes on its least (largest) value.

We give this solution recipe the short name of the *Lagrange principle*.

The theorems just formulated imply that *if* (in the problem without in-equalities) *the set of admissible points is bounded and all functions* $f_0, \ldots, f_m$ *as well as all partial derivatives* $\partial f_i / \partial x_j$, $i = 0, \ldots, m$, $j = 1, \ldots, n$, *are continuous, then the rule just given yields a solution of the problem.*

There is no need to study the art of differentiating functions of many variables; after all, the matter reduces to differentiating functions of one variable. One could proceed directly to the solution of problems, but we prefer to devote more time to discussion.

2. At this point, just as in the previous story, we will clarify some of the points discussed in the first section.

In the multidimensional case there is no need to clarify Fermat's theorem further, for it follows trivially from the one-dimensional case. Indeed, let the function $f_0(x_1, \ldots, x_n)$ have a local extremum at the point $(\hat{x}_1, \ldots, \hat{x}_n)$. Then the function $g_j(x)$ (similar to the one defined in §1) must have a minimum at 0. But then, according to the one-dimensional version of Fermat's theorem discussed in detail in the previous story,

$$g_j'(0) = 0.$$

From the definition of the jth partial derivative alone we see that

$$g_j'(0) = \frac{\partial f_0(\hat{x})}{\partial x_j}.$$

Juxtaposition of these two equalities for the different values of j yields Fermat's theorem

$$\frac{\partial f_0(\hat{x})}{\partial x_j} = \frac{\partial f_0(\hat{x}_1, \ldots, \hat{x}_n)}{\partial x_j} = 0, \qquad j = 1, \ldots, n.$$

It remains to examine the Lagrange multiplier rule. We will begin the simplest situation, namely, when $n = 2$, $m = 1$, and the constraint is given by *a linear relation*. In other words, we consider the problem

$$f_0(x_1, x_2) \to \max, \qquad f_1(x_1, x_2) = a_1 x_1 + a_2 x_2 - b = 0.$$

The graph of $y = f_0(x_1, x_2)$ can be thought of as the landscape of a mountainous region (recall what was said in this connection in the tenth story). The relation $f_1(x_1, x_2) = a_1 x_1 + a_2 x_2 - b = 0$ determines a line in the plane. Think of an electric transmission line being built in a mountainous region along a path whose representation on the map is a line.

QUESTION. Where is the highest point of the route of the transmission line?

Recall how mountains are represented on a map. The map shows level lines, that is, curves connecting points at the same altitude. Now think of the mutual disposition of the route (on the map) of the transmission line and the level line of the mountain at the highest point of the route.

Clearly, *at the point in question the route cannot cross the level line.* For if it did, then it would be crossing from lower to higher values of the altitude. Hence an intersection point cannot be a point of maximal height. We conclude that *at the maximum point the route must be tangent to the level line.*

Now we consider the function $f_1(x_1, x_2) = a_1 x_1 + a_2 x_2 - b$. Its partial derivatives are

$$\frac{\partial f_1}{\partial x_1} = a_1, \qquad \frac{\partial f_1}{\partial x_2} = a_2.$$

The vector (a_1, a_2) is perpendicular to the route and to every level line of the function $f_1(x_1, x_2)$. This is clear from the geometric sense of the equation $a_1 x_1 + a_2 x_2 = b$. But this turns out to be always true. Specifically, if $f(x_1, x_2)$ is a continuous function with continuous partial derivatives, then *the vector $(\partial f(\overline{x})/\partial x_1, \partial f(\overline{x})/\partial f_2)$ is perpendicular to the tangent at $\overline{x}$ to the level line $f(x) = f(\overline{x})$.* As noted earlier, *the route is tangent to the level line at the maximum point of the altitude.* All this implies that both vectors $(\partial f_0(\hat{x})/\partial x_1, \partial f_0(\hat{x})/\partial x_2)$ and (a_1, a_2) are perpendicular to the route and, therefore, proportional, that is,

$$\frac{\partial f_0(\hat{x})}{\partial x_1} + \lambda a_1 = 0, \qquad \frac{\partial f_0(\hat{x})}{\partial x_2} + \lambda a_2 = 0,$$

or

$$\frac{\partial \mathscr{L}(\hat{x}, 1, \lambda)}{\partial x_1} = 0, \qquad \frac{\partial \mathscr{L}(\hat{x}, 1, \lambda)}{\partial x_2} = 0,$$

where

$$\mathscr{L}(x, \lambda_0, \lambda) = \lambda_0 f_0(x) + \lambda f_1(x).$$

In this special case we have arrived at the Lagrange multiplier rule.

Now suppose that the function f_1 is not necessarily affine but has continuous partial derivatives. We will again consider the families of level lines

$$f_0(x_1, x_2) = c_0, \qquad f_1(x_1, x_2) = c_1.$$

We assume that through each point of a certain part of the plane there passes just one curve from each family. Suppose that the problem has a solution at $\hat{x} = (\hat{x}_1, \hat{x}_2)$. Then all points of the curve l_0 given by the equation

$$f_0(x_1, x_2) = \hat{c}_0 =: f_0(\hat{x}_1, \hat{x}_2),$$

must lie "on one side" of the curve l_1 given by the equation

$$f_1(x_1, x_2) = \hat{c}_1 =: f_1(\hat{x}_1, \hat{x}_2),$$

that is, these curves don't intersect but are tangent to one another. What we have tried to make clear is that if the function f_0 takes on an extremal value at the point $\hat{x}$ on the curve l_1, then the curve l_0 is tangent to the curve l_1.

Recall once more that the vector $(\partial f_0(\hat{x})/\partial x_1, \partial f_0(\hat{x})/\partial x_2)$ is perpendicular to the curve l_0 and the vector $(\partial f_1(\hat{x})/\partial x_1, \partial f_1(\hat{x})/\partial x_2)$ is perpendicular to the curve l_1. If these curves are tangent to one another, then the vectors in question are proportional, as is claimed in the Lagrange multiplier rule (with $\lambda_0 = 1$).

In this book we won't be able to prove the Lagrange principle (that is, the Lagrange multiplier rule). However, some explanations bearing on its proof will be given in the fourteenth story.

At the beginning of the present story we formulated the general problem involving equalities as well as inequalities. But later we dealt only with the case of equalities. It is natural to ask what changes in the Lagrange principle in the general case. Let's turn to problem (p) posed at the beginning of this story and assume that all functions $f_0(x), \ldots, f_m(x)$ satisfy the conditions of the theorem on the Lagrange multiplier rule formulated earlier. If an admissible point $\hat{x}$ yields a local minimum in problem (p) then there are numbers $\lambda_0, \lambda_1, \ldots, \lambda_m$, not all zero, such that

$$\frac{\partial \mathcal{L}(\hat{x}, \lambda)}{\partial x_j} = 0, \qquad j = 1, \ldots, n,$$

$$(\hat{x} = (\hat{x}_1, \ldots, \hat{x}_n), \qquad \lambda = (\lambda_0, \lambda_1, \ldots, \lambda_m)),$$

the multipliers at the functional and at the inequalities satisfy the nonnegativeness conditions $\lambda_0 \geq 0$, $\lambda_{m'+1} \geq 0, \ldots, \lambda_m \geq 0$, and the following conditions (called *supplementary slackness conditions*) hold:

$$\lambda_j f_j(\hat{x}) = 0, \qquad j = m' + 1, \ldots, m.$$

The supplementary slackness conditions mean that a Lagrange multiplier λ_j can be different from zero only at an "active" constraint, when at an extremum point an inequality constraint is actually an equality: $f_j(\hat{x}) = 0$, $j = m' + 1, \ldots, m$.

What has changed? Firstly, "extremum" changed to "minimum." When inequalities are present, the type of extremum being considered is not irrelevant. Before we can apply the result just formulated in the case of a maximum problem or a problem involving inequalities of the form $f_j \geq 0$, we must change the problem to one of the form (p) by possibly changing some of the f_j, $j = 0, m' + 1, \ldots, m$, to $-f_j$. We've added, secondly, the nonnegativeness conditions, and thirdly, the supplementary slackness conditions.

When solving problems in this book we won't need to use the theorem just formulated. But its role, and the role of similar theorems, is very significant. In the first story we hinted that the old methods have turned out to be inadequate for the solution of many economic problems. Now it is possible to be more concrete.

Recall the transportation problem discussed twice before. Its formalization involves inequalities, and, in fact, it is difficult to formalize the problem without them. Earlier, one did not consider problems with inequalities. The just-stated addition to the Lagrange multiplier rule turned up twenty-odd years ago (not two hundred and fifty years ago). In addition, it became clear that, in the case of economic problems, the minimized functions and conditions are convex, even linear. This made it necessary to study convex functions and convex extremal problems. We, too, will now turn our attention to these issues.

3. Convex functions and convex problems. We've already touched on convex functions of one variable. Convex functions of many variables are defined in an entirely similar manner. Thus a function $f(x) = f(x_1, \ldots, x_n)$ is said to be *convex* if for arbitrary points x and x' and any α, $0 \le \alpha \le 1$, the Jensen inequality

$$f(\alpha x + (1 - \alpha)x') \le \alpha f(x) + (1 - \alpha)f(x')$$

holds.

Examples of convex functions are, first, linear and affine functions. We note also the *distance function* from a point to the origin

$$f(x_1, \ldots, x_n) = \sqrt{x_1^2 + \cdots + x_n^2}.$$

A function $y = f(x)$ is said to be *strictly convex* if in the Jensen inequality, for $x \ne x'$ and $0 < \alpha < 1$, we have the strict inequality

$$f(\alpha x + (1 - \alpha)x') < \alpha f(x) + (1 - \alpha)f(x').$$

The functions $y = x^2$, $y = \sqrt{h^2 + x^2}$, and $x_1^2 + x_2^2$ are strictly convex, whereas the function $y = |x|$ and the distance function are not. It is easy to see that if a strictly convex function $y = f(x)$ attains its minimum at a point $\hat{x}$ then this minimum is unique.

We noted that not all functions are differentiable and twice we discussed the example $y = |x|$. This function is convex and not differentiable at zero. The distance function is also differentiable everywhere except at the origin. However, if a convex function is differentiable and its derivative vanishes at some point, then the function attains its absolute extremum at this point. It is also true that the graph of a convex function always lies above any of its tangent planes.

This fact is of great importance: for convex differentiable functions Fermat's theorem is a sufficient condition for an extremum! This is one of the reasons why the theory of convex extremal problems is so complete.

13

More Problem Solving

Here we intend to fulfill our earlier (maybe rash) promise. Our aim is to solve again all problems from Part One "in the same, standard, you might even say routine, way," using the same (simple) method, namely, the Lagrange principle, or, in special cases, Fermat's theorem. (We won't be able to do this is three cases: the classical isoperimetric problem, the brachistochrone problem, and Newton's problem. To solve these problems routinely, we'll have to invest more effort.)

Our standard approach will involve four stages: (1) formalization; (2) the use of the Lagrange principle or Fermat's theorem; (3) solution of the corresponding equations and location of the critical or stationary points; and (4) selection of the required points and discussion of the answer.

Let's begin with the problems that reduce to finding extrema of functions of one variable.

1. Euclid's problem on the parallelogram of maximal area inscribed in a triangle (fourth story)

$1°$ *Formalization.* Let's turn again to Figure 4.1 on p. 28. As before, let H denote the height of the triangle ABC and b the length of AC. Let x' be the length of AF'. Then $0 \leq x \leq b$. Let $h = h(x)$ denote the height of the triangle $BD'E'$. The similarity of the triangles $BD'E'$ and ABC ($D'E' \| AC$) implies that $h(x)/H = x/b$. The area of the parallelogram $AD'E'F'$ is equal to $(H - h(x))x = H(b - x)x/b$. In sum, we arrive at the following formalization:

$$(\text{p}_1) \qquad f_0(x) = \frac{Hx(b - x)}{b} \to \max, \qquad 0 \leq x \leq b.$$

$2°$ *Necessary condition.* $f_0'(x) = 0$.

$3°$ *Finding the critical points.* The stationary points: $f_0'(x) = (H(bx - x^2)/b)' = (b - 2x)H/b$, i.e. $f_0'(x) = 0$ only at the point $b/2$. The critical points are 0, $b/2$, and b.

$4°$ *Discussion.* The function f_0 is continuous, differentiable everywhere, and is considered on a finite interval. This means that the solution is among the critical points. Sorting out the critical points:

$$f_0(0) = f_0(b) = 0, \qquad f_0(b/2) > 0.$$

It follows that the solution of (p_1) is $b/2$. Answer: *The required parallelogram ADEF is characterized by the fact that the point F is the midpoint of the segment* $[AC]$. The same fact was established by Euclid.

2. Archimedes' problem on the spherical segment of largest volume among the isoepiphanic ones (fourth story)

$1°$ *Formalization.* Let R be the radius of the sphere and h the height of the spherical segment. It is well-known that the volume of a spherical segment is $\pi h^2(R - h/3)$, and its lateral surface is $2\pi R h$. Since the area of the lateral surface is given, $2\pi R h = a$, $R = a/2\pi h$. Substituting this value for R in the volume formula and noting that $h \le 2R = a/\pi h$, we obtain the following formalization:

$$(\text{p}_2) \qquad f_0(h) = \frac{ha}{2} - \frac{\pi h^3}{3} \to \max, \qquad 0 \le h \le \sqrt{a/\pi}.$$

$2°$ *Necessary condition:* $f_0'(h) = 0$.

$3°$ *Finding the critical points.* The stationary points: $f_0'(h) = (ha/2 - \pi h^3/3)' = a/2 - \pi h^2$, i.e. $f_0'(h) = 0$ only if $h = \sqrt{a/2\pi}$. The critical points: 0, $\sqrt{a/2\pi}$, $\sqrt{a/\pi}$.

$4°$ *Discussion.* The function f_0 is continuous and differentiable everywhere, and is considered on a finite interval. Hence the solution is among the critical points. Sorting out the critical points: $f_0(0) = 0$, $f_0(\sqrt{a/2\pi}) = \sqrt{2}a^{3/2}/6\sqrt{\pi}$, $f_0(\sqrt{a/\pi}) = a^{3/2}/6\sqrt{\pi}$. The point $\sqrt{a/2\pi}$ yields the maximal value. This is the solution. Since $a = 2\pi R h$, we obtain $h = R$. Answer: *The required spherical segment is a hemisphere—its height equals the radius.* The same result was established by Archimedes.

3. The problem of least area (fourth story)

$1°$ *Formalization.* Let's turn once again to Figure 4.6 on p. 33. We draw a line through the point M parallel to AB and denote by N its point of intersection with AC. Let E' be a point on the ray NG, $a = |AN|$, $x = |NE'|$, and D' the point of intersection of the ray AB and the line $E'M$. Since $MN \| AB$, the triangles $ME'N$ and $AD'E'$ are similar. This means that the ratio of their areas is the same as the ratio of the squares of the lengths of the segments $[NE']$ and $[AE']$. But the area of the triangle $ME'N$ is $xh/2$, where h is the altitude from M to AC. Hence the area of the triangle $AD'E'$ is $h(x + a)^2/2x$, and we arrive at the following formalization:

$$(\text{p}_3) \qquad f_0(x) = \frac{(a + x)^2}{x} \to \min, \qquad x > 0.$$

$2°$ *Necessary condition.* $f_0'(x) = 0$.

$3°$ *Finding the stationary points.*

$$f_0'(x) = \left(\frac{(a + x)^2}{x} \right)' = \left(\frac{a^2}{x} + 2a + x \right)' = -\frac{a^2}{x^2} + 1,$$

that is $f_0'(x) = 0$ only for $x = a$.

$4°$ *Discussion.* The function f_0 satisfies the requirements of the corollary to Weierstrass' theorem in the eleventh story. This means that *problem* (p_3) *is solvable.* Since f_0 is differentiable for $x > 0$, Fermat's theorem implies that *the solution must be a stationary point.* But the stationary point is *unique.* This means that it is the solution. Answer: *the required point E is at a distance of $2a$ from A.* It follows that the point M halves the segment DE (in view of the similarity of the triangles ENM and EAD). We obtained this result before by geometric means.

4. Heron's problem. We've encountered this problem many times—in the first, second, tenth, and eleventh stories. It's time now to solve it the standard way.

$1°$ *Formalization.* This was carried out in the tenth story:

$$(\text{p}_4) \qquad f_0(x) = \sqrt{a^2 + x^2} + \sqrt{b^2 + (d - x)^2} \to \min.$$

$2°$ *Necessary condition.* $f_0'(x) = 0$.

$3°$ *Finding the stationary points.* Using the theorem on differentiating a function of a function, we obtain:

$$\left(\sqrt{a^2 + x^2} \right)' = x / \sqrt{a^2 + x^2},$$

(this was carried out in detail in the tenth story) and

$$\left(\sqrt{b^2 + (d-x)^2}\right)' = -\frac{d-x}{\sqrt{a^2 + (d-x)^2}}.$$

Hence

(1) $$\frac{x}{\sqrt{a^2 + x^2}} = \frac{d-x}{\sqrt{b^2 + (d-x)^2}}.$$

To solve this equation we square, invert, and subtract 1 from both sides. This yields the relation $(a/x)^2 = (b/(d-x))^2$. After eliminating the extraneous root, we obtain the equality $x/a = (d-x)/b$ (which coincides with what was said in the first story; see Figure 1.1). We denote the solution of the latter equation by $\hat{x}$.

$4°$ *Discussion.* As a sum of two convex functions, the function $f_0(x)$ is convex. It is also smooth. From what was said at the end of the previous story, we know that $\hat{x}$ is a solution of the problem. Since $f_0(x)$ is strictly convex, the solution $\hat{x}$ is unique.

Let's take another look at the very first figure in the book. The quantity $\hat{x}/\sqrt{a^2 + \hat{x}^2}$ is equal to $\sin\varphi_1$, and $(d-\hat{x})/\sqrt{b^2 + (d-\hat{x})^2}$ to $\sin\varphi_2$. From (1) it follows that $\sin\varphi_1 = \sin\varphi_2$, that is $\varphi_1 = \varphi_2$.

Answer: *What characterizes the solution of Heron's problem is the equality of the angles of incidence and reflection*—a fact we established at the very beginning of this book.

In the first story we stated problems 1 and 2, which are close to Heron's problem and easily reduce to it. In problem 1 show that a solution exists. If it does not coincide with the vertex of the angle, then it is the solution of Heron's problem for B and C and the side of the angle on which the point A should be. This means that the angles of incidence and reflection at A must be equal. A similar assertion holds for the point B. This leads to the required construction. Problem 2 is solved in the same way. Therefore there is no need to solve these problems formally.

5. Snel's problem on the law of refraction of light (third story). Let's solve this problem the standard way, as Leibniz was the first to do.

$1°$ *Formalization.* We take the line separating the two media as the x-axis and the line through A perpendicular to it as the y-axis (see Figure 3.2 on page 21). Let the coordinates of A and B be $A = (0, a)$, $B = (d, -b)$. Let D' be a point on the x-axis with coordinates $(x, 0)$. The time it takes for light to traverse the path $AD'B$ is $\sqrt{a^2 + x^2}/v_1 + \sqrt{b^2 + (d-x)^2}/v_2$.

This leads to the following unconstrained problem:

$$(p_5) \qquad f_0(x) = \frac{\sqrt{a^2 + x^2}}{v_1} + \frac{\sqrt{b^2 + (d - x)^2}}{v_2} \to \min .$$

$2°$ *Necessary condition.* $f_0'(x) = 0$.

$3°$ *Finding the stationary points.*

(1)
$$f_0'(x) = \left(\frac{\sqrt{a^2 + x^2}}{v_1} \right)' + \left(\frac{\sqrt{b^2 + (d - x)^2}}{v_2} \right)'$$

$$= \frac{x}{v_1 \sqrt{a^2 + x^2}} - \frac{d - x}{v_2 \sqrt{b^2 + (d - x)^2}} = 0 .$$

(Concerning differentiation, see the previous story.) In view of the mono-tonicity of the functions $x/v_1\sqrt{a^2 + x^2}$ and $(d - x)/v_2\sqrt{b^2 + (d - x)^2}$ we see that the equation $f_0'(x) = 0$ has a unique solution $\hat{x}$.

$4°$ *Discussion.* In view of the convexity of the function $f_0(x)$ Fermat's theorem is a sufficient condition for an extremum. Hence $\hat{x}$ is a solution. The strict convexity of $f_0(x)$ implies that this solution is unique. Figure 3.2 on p. 21 and relation (1) imply the equality

$$\frac{\sin \alpha_1}{v_1} = \frac{\sin \alpha_2}{v_2}$$

that expresses Snel's law. Answer: *the solution of Snel's problem is charac-terized by the equality of the ratio of the sines of the angles of incidence and refraction and the ratio of the velocities in the first and second media*—a fact we established in the third story.

6. Kepler's planimetric problem on a rectangle of maximal area inscribed in a circle. We discussed this problem in the fifth and tenth stories; it was formalized in the tenth story.

$1°$ *Formalization.*

$$(p_6) \qquad f_0(x) = x\sqrt{1 - x^2} \to \max, \qquad 0 \le x \le 1 .$$

$2°$ *Necessary condition.* $f_0'(x) = 0$.

$3°$ *Finding the critical points.* The stationary points: $f_0'(x) = (x\sqrt{1 - x^2})'$ $= \sqrt{1 - x^2} + x(\sqrt{1 - x^2})' = \sqrt{1 - x^2} - x^2/\sqrt{1 - x^2} = 0 \Leftrightarrow 2x^2 = 1 \Rightarrow x = \sqrt{2}/2$. Thus there are three critical points: 0, 1, and $\sqrt{2}/2$.

$4°$ *Discussion.* The function f_0 is continuous on $[0, 1]$ and differentiable in $(0, 1)$. This means that the solution is among the critical points. Sorting out the critical points: $f_0(0) = f_0(1) = 0$, $f_0(\sqrt{2}/2) = \sqrt{2}/4$. Hence $\sqrt{2}/2$ is the solution to (p_6). In this case, as implied by the formalization, the rectangle is square. Answer: *The largest rectangle inscribed in a circle is a square.*

7. Kepler's problem on an inscribed cylinder. We talked about this problem in the sixth story.

$1°$ *Formalization.* Let R be the radius of the sphere. Let x be half the height of the cylinder. Then $0 \leq x \leq R$. The base radius of the cylinder is $\sqrt{R^2 - x^2}$ and its volume is $\pi r^2 h = 2\pi(R^2 - x^2)x$. Hence the formalization:

(p_7) $\qquad\qquad f_0(x) = 2\pi(R^2 - x^2)x \to \max, \qquad 0 \leq x \leq R.$

(Actually, this problem was formalized in the sixth story.)

$2°$ *Necessary condition.* $f_0'(x) = 0$.

$3°$ *Finding the critical points.* The stationary points are

$$f_0'(x) = (2\pi(R^2 - x^2)x)' = 2\pi(R^2 x - x^3)'$$
$$= 2\pi(R^2 - 3x^2) = 0 \Leftrightarrow x_1 = R/\sqrt{3}, \ x_2 = -R/\sqrt{3}.$$

The second root is unsuitable $(x_2 < 0)$. Hence there are three critical points: $0, R/\sqrt{3}$, and R.

$4°$ *Discussion.* The function f_0 is continuous and differentiable everywhere. This means that the solution is among the critical points. Since $f_0(0) = f_0(R) = 0$, the solution is $R/\sqrt{3}$. Hence the radius of the maximal cylinder is $\sqrt{R^2 - R^2/3} = R\sqrt{2/3}$. Answer: *The ratio of the height of the extremal cylinder to the base diameter is $\sqrt{2}$.* This is the fact established by Kepler.

We will now solve some algebraic problems.

8. Tartaglia's problem (fifth story).

$1°$ *Formalization.* Let x be the smaller number. Then $0 \leq x \leq 4$ and the larger number is $8 - x$. Their difference is $8 - 2x$. In sum,

(p_8) $\qquad\qquad f_0(x) = x(8 - x)(8 - 2x) \to \max, \qquad 0 \leq x \leq 4.$

$2°$ *Necessary condition.* $f_0'(x) = 0$.

3° *Finding the critical points.* The stationary points are

$$f_0'(x) = (x(8-x)(8-2x))' = (2x^3 - 24x^2 + 64x)'$$
$$= 6x^2 - 48x + 64 = 0 \Leftrightarrow x_1 = 4 - 4/\sqrt{3}, \ x_2 = 4 + 4/\sqrt{3}.$$

The second root is unsuitable $(x_2 > 4)$. Thus there are three critical points: 0, 4, and $4 - 4/\sqrt{3}$.

4° *Discussion.* The function f_0 is continuous and differentiable everywhere, and is considered on a finite interval. This means that the solution is among the critical points. Since $f_0(0) = f_0(4) = 0$ and $f_0(4 - 4/\sqrt{3}) > 0$, it follows that $4 - 4/\sqrt{3}$ is the solution of (p_8). Answer: *The larger number is $4 + 4/\sqrt{3}$ and the smaller one is $4 - 4/\sqrt{3}$.* This fact was established by Tartaglia.

9. The inequality of the arithmetic-geometric means (fifth story). We consider an auxiliary extremal problem:

$$
\begin{aligned}
& f_0(x) = x_1 \cdot x_2 \cdots x_n \to \max, \\
(p_9) \quad & f_1(x) = x_1 + x_2 + \cdots + x_n = 1, \\
& f_i(x) = x_{i-1} \geq 0, \quad i = 2, 3, \ldots, n+1 \quad (x = (x_1, \ldots, x_n)).
\end{aligned}
$$

The functions f_i and their partial derivatives are continuous. Since $0 \leq x_k \leq 1$ for all k, the set of admissible points is bounded. Hence Weierstrass' theorem implies the existence of a solution $\hat{x} = (\hat{x}_1, \ldots, \hat{x}_n)$. Of course, $\hat{x}_k \neq 0$; otherwise $f_0(\hat{x}) = 0$, at a time when there exist admissible elements with $f_0(x) > 0$.

Of course, $\hat{x}$ will also be a local maximum in problem (p_9). Since, as was just shown, $\hat{x}_k > 0$, it will also be a local maximum in the problem without inequalities.

1° *Formalization.*

$$(p') \qquad\qquad f_0(x) \to \max, \qquad f_1(x) = 1.$$

The Lagrange function for (p') is $\mathscr{L}(x, \lambda_0, \lambda_1) = \lambda_0 f_0(x) + \lambda_1 f_1(x)$.

2° *Necessary condition.* This condition is Lagrange multiplier rule:

$$\frac{\partial \mathscr{L}}{\partial x_k} = 0, \qquad k = 1, \ldots, n.$$

3° *Finding the stationary points.* Let A denote the product $\hat{x}_1 \cdots \hat{x}_n$. Then

$$\frac{\partial \mathscr{L}}{\partial x_k}(\hat{x}, \lambda_0, \lambda_1) = 0 \Rightarrow \frac{\lambda_0 A}{\hat{x}_k} + \lambda_1 = 0.$$

The assumption $\lambda_1 = 0$ would imply the untenable conclusion that both multipliers λ_0 and λ_1 are zero. Thus $\hat{x}_k = -\lambda_0 A/\lambda_1$, that is, $\hat{x}_1 = \cdots = \hat{x}, = 1/n$, (since $\hat{x}_1 + \cdots + \hat{x}_n = 1$).

$4°$ *Discussion.* Since the stationary point in (p$'$) is unique, it yields the solution of the problem.

We can now prove the required inequality. Let $a_1, \ldots, a_n$ be arbitrary nonnegative numbers. Put $S = a_1 + \cdots + a_n$ and $x_j = a_j/S$. Then $x_1 + \cdots + x_n = 1$ and, by what was just proved.

$$\frac{a_1 \cdot a_2 \cdots a_n}{S^n} = x_1 \cdot x_2 \cdots x_n \le \frac{1}{n^n}$$

$$\Rightarrow a_1 \cdot a_2 \cdots a_n \le \left(\frac{a_1 + a_2 + \cdots + a_n}{n}\right)^n,$$

which is the required result.

10. The inequality of the arithmetic-quadratic means (fifth story).

$1°$ *Formalization.*

(p_{10})
$$f_0(x) = x_1 + \cdots + x_n \to \max,$$
$$f_1(x) = x_1^2 + \cdots + x_n^2 = 1 \qquad (x = (x_1, \ldots, x_n)).$$

The functions f_0 and f_1 and their partial derivatives are continuous. Since $-1 \le x_k \le 1$, $k = 1, \ldots, n$, the set of admissible points is bounded. This means that a solution exists, and we can use the Lagrange principle. The Lagrange function is $\mathscr{L}(x, \lambda_0, \lambda_1) = \lambda_0 f_0(x) + \lambda_1 f_1(x)$.

$2°$ *Necessary condition.*

$$\frac{\partial \mathscr{L}}{\partial x_j} = 0, \qquad j = 1, \ldots, n.$$

$3°$ *Finding the stationary points.*

$$\frac{\partial \mathscr{L}}{\partial x_j}(\hat{x}, \lambda_0, \lambda_1) = \lambda_0 + 2\lambda_1 \hat{x}_j = 0.$$

The assumption that $\lambda_1 = 0$ leads to the untenable conclusion that both Lagrange multipliers are zero. Hence $\hat{x}_j = -\lambda_0/2\lambda_1$, that is $\hat{x}_1 = \cdots = \hat{x}_n = 1/\sqrt{n}$ (for $\hat{x}_1^2 + \cdots + \hat{x}_n^2 = 1$).

$4°$ *Discussion.* Since the stationary point is unique, it is the solution of the problem.

Now we can prove the required inequality. Let $a_1, \ldots, a_n$ be arbitrary numbers. Put $S = (a_1^2 + \cdots + a_n^2)^{1/2}$ and $x_j = a_j/S$. Then $x_1^2 + \cdots + x_n^2 = 1$

and, in view of what was just proved,

$$\frac{a_1 + \cdots + a_n}{S} = x_1 + \cdots + x_n \le \frac{n}{\sqrt{n}} = \sqrt{n} \Rightarrow$$

$$\frac{a_1 + \cdots + a_n}{n} \le \frac{S}{\sqrt{n}} = \left(\frac{a_1^2 + \cdots + a_n^2}{n}\right)^{1/2},$$

which is the required result.

This also proves the assertion that (for nonnegative a_i)

$$\sqrt[n]{a_1 \cdots a_n} \le \left(\frac{a_1^2 + \cdots + a_n^2}{n}\right)^{1/2},$$

and thus yields a solution of Kepler's planimetric problem and of one of the stereometric Kepler problems discussed in the fifth story.

Proceeding analogously, one can prove the following general inequalities for means. Let $a_1, \ldots, a_n$ be nonnegative numbers. Put

$$S_p = \left(\frac{a_1^p + \cdots + a_n^p}{n}\right)^{1/p}, \qquad \sum_p = (a_1^p + \cdots + a_n^p).$$

Then

(1) $$S_p \le S_q \quad \text{if } p \le q,$$

(2) $$\sum_p \le \sum_q \quad \text{if } p \ge q.$$

Earlier we proved (1) for $p = 1$, $q = 2$.

11. The Cauchy-Bunyakovskiĭ and Hölder inequalities (fifth story). Let $a_1, \ldots, a_n$ be fixed numbers not all zero.

$1°$ *Formalization.* We consider the extremal problem

$$(\text{p}_{11}) \qquad f_0(x) = a_1 x_1 + \cdots + a_n x_n \to \max,$$
$$f_1(x) = x_1^2 + \cdots + x_n^2 = B^2 \qquad (x = (x_1, \ldots, x_n)).$$

The functions f_0 and f_1 and their partial derivatives are continuous. Since $-B \le x_j \le B$, $j = 1, \ldots, n$, the set of admissible points is bounded. Hence a solution exists, and we can apply the Lagrange principle. The Lagrange function is $\mathcal{L}(x, \lambda_0, \lambda_1) = \lambda_0 f_0(x) + \lambda_1 f_1(x)$. We put

$$A = (a_1^2 + \cdots + a_n^2)^{1/2}.$$

$2°$ *Necessary condition.*

$$\frac{\partial \mathcal{L}}{\partial x_j} = 0.$$

3° *Finding the stationary points.*

$$\frac{\partial \mathscr{L}}{\partial x_j}(\hat{x}, \lambda_0, \lambda_1) = \lambda_0 a_j + 2\lambda_1 \hat{x}_j, \qquad j = 1, \dots, n.$$

The possibility $\lambda_1 = 0$ implies the untenable conclusion that both Lagrange multipliers are zero. We have

$$\hat{x}_j = \lambda_0 a_j / 2\lambda_j = C a_j, \quad j = 1, \dots, n.$$

Since $\hat{x}_1^2 + \cdots + \hat{x}_n^2 = B^2$, it follows that

$$C^2(a_1^2 + \cdots + a_n^2) = B^2 \Rightarrow C = \pm B/A.$$

4° *Discussion.* There are just two stationary points. The solution is the point corresponding to the plus sign: $\hat{x}_j = Ba_j/A$.

Now let $b_1, \dots, b_n$ be any n numbers and $B^2 = b_1^2 + \cdots + b_n^2$. By what has just been proved,

$$a_1 b_1 + \cdots + a_n b_n \le a_1 \frac{Ba_1}{A} + \cdots + a_n \frac{Ba_n}{A} = AB = (a_1^2 + \cdots + a_n^2)^{1/2}(b_1^2 + \cdots + b_n^2)^{1/2},$$

which is *the Cauchy-Bunyakovskiĭ inequality.*

The *Hölder inequality* is proved in the same way. We set down the necessary computations without comment.

1° *Formalization.*

$(\mathbf{p}'_{11}) \qquad f_0(x) = a_1 x_1 + \cdots + a_n x_n \to \max,$

$$f_1(x) = |x_1|^p + \cdots + |x_n|^p = B^p, \qquad (a_i > 0, \; x = (x_1, \dots, x_n)).$$

The Lagrange function is $\mathscr{L} = \lambda_0 f_0(x) + \lambda_1 f_1(x)$. We set $A = (a_1^{p'} + \cdots + a_n^{p'})$, $p'^{-1} + p^{-1} = 1$.

2° *Necessary condition.*

$$\frac{\partial \mathscr{L}}{\partial x_j} = 0, \qquad j = 1, \dots, n.$$

3° *Finding the stationary points.*

$$\lambda_0 a_j + p\lambda_1 |x_j|^{p-1} \operatorname{sign} x_j = 0 \Rightarrow \hat{x}_j = C a_j^{p'-1}, \qquad C = \pm B/A^{p'/p}.$$

4°. Solution: $\hat{x}_j = Ba_j^{p'-1}/A^{p'/p}, \; j = 1, \dots, n.$

Now let $b_1, \dots, b_n$ be any n nonnegative numbers. By what was just proved,

$$a_1 b_1 + \cdots + a_n b_n \le a_1 \frac{B}{A^{p'/p}} a_1^{p'-1} + \cdots + a_n \frac{B}{A^{p'/p}} a_n^{p'-1} = BA^{p'(1-1/p)},$$

which is the *Hölder inequality.*

Now we again turn to geometry.

12. The Steiner problem (fourth story). Let the coordinates of the three given points be $A(a_1, a_2)$, $B(b_1, b_2)$, and $C(c_1, c_2)$. Let D be a point with coordinates (x_1, x_2). Then the sum of the distances from D to A, B, and C is

$$f_0(x_1, x_2) = \sqrt{(x_1 - a_1)^2 + (x_2 - a_2)^2} + \sqrt{(x_1 - b_1)^2 + (x_2 - b_2)^2}$$
$$+ \sqrt{(x_1 - c_1)^2 + (x_2 - c_2)^2}.$$

This leads to an unconstrained problem.

$1°$ *Formalization.*

(p_{12}) $\qquad\qquad\qquad\qquad$ $f_0(x_1, x_2) \to \min$.

Note that if $x_1^2 + x_2^2$ is large, then D is located far from A, B and C, and so the sum of its distances from the points A, B, and C is also large. Hence $x_1^2 + x_2^2 \to \infty$ implies $f_0(x) \to \infty$. But then we can use the corollary to Weierstrass' theorem in the previous story and conclude that problem (p_{12}) has a solution $\hat{x} = (\hat{x}_1, \hat{x}_2)$.

It is easy to see that the partial derivatives of the function f_0 exist and are continuous at all points x other than A, B, and C.

$2°$ *Necessary (and, in view of the convexity of f_0, sufficient) condition—the Fermat theorem.* If $\hat{x} \neq A, B, C$, then

$$\frac{\partial f_0(\hat{x})}{\partial x_1} = \frac{\partial f_0(\hat{x})}{\partial x_2} = 0.$$

$3°$ *Finding the stationary points.*

$$\frac{\partial f_0}{\partial x_1}(\hat{x}_1, \hat{x}_2) = \frac{\hat{x}_1 - a_1}{|\overrightarrow{DA}|} + \frac{\hat{x}_1 - b_1}{|\overrightarrow{DB}|} + \frac{\hat{x}_1 - c_1}{|\overrightarrow{DB}|},$$

$$\frac{\partial f_0}{\partial x_2}(\hat{x}_1, \hat{x}_2) = \frac{\hat{x}_2 - a_2}{|\overrightarrow{DA}|} + \frac{\hat{x}_2 - b_2}{|\overrightarrow{DB}|} + \frac{\hat{x}_2 - c_2}{|\overrightarrow{DC}|}.$$

$4°$ *Discussion.* We clarify the geometric sense of the relations just set down. They state that the sum of the unit vectors

$$e_1 = \frac{\overrightarrow{DA}}{|\overrightarrow{DA}|}, e_2 = \frac{\overrightarrow{DB}}{|\overrightarrow{DB}|}, e_3 = \frac{\overrightarrow{DC}}{|\overrightarrow{DC}|}$$

is zero. But then one can make out of them an equilateral triangle, that is, each of the angles $A\hat{D}B$, $B\hat{D}C$, and $C\hat{D}A$ is $120°$. *This means that if the solution does not coincide with one of the vertices of the triangle ABC, then $\hat{D}$ is a point from which each side is seen at an angle of $120°$.* Hence

$\hat{D}$ is just *the Torricelli point* talked about in the fourth story (where we also learned how to construct it). If the obtuse angle in a triangle is $\geq 120°$, then there is no point from which all sides can be seen at an angle of $120°$. This means that $\hat{x}$ must coincide with one of the vertices, namely, the vertex of the obtuse angle, because the larger side lies opposite the larger angle. (Let C be the vertex of the obtuse angle. Then, using the natural symbols for the sides, $c > a$ and $c > b$. This means that $a + b < a + c$ and $a + b < b + c$, that is, $a + b$ is the least of the three sums.)

Answer. *If all angles in the triangle are* $< 120°$, *then the required point is its Torricelli point. If one of the angles is* $\geq 120°$, *then the required point coincides with the vertex of this angle.* This is just the answer we derived in the fourth story.

In the first story we formulated Problem 4, which is very close to Steiner's problem. The answer to this problem was given in the fourth story. In the part pertaining to a convex quadrilateral, the answer follows immediately from the sufficiency of Fermat's theorem for an unconstrained convex problem. In the nonconvex case one proves, just as in Steiner's problem, that a solution exists. Then one verifies that if the point is different from a vertex, then the necessary condition cannot be fulfilled.

13. The problem of the least perimeter (fourth story). Let's turn to Figure 4.7. Consider lines through M parallel to AB and AC and denote by N and P their respective points of intersection with AC and AB. We denote $|AN|$ by a, $|AP|$ by b, the angle BAC by α, a segment through M by $[D'E']$ (D' on AB, E' on AC), $|NE'|$ by x, $|PD'|$ by y, the angle $AD'E'$ by ψ, and the angle $D'E'C$ by φ.

The similarity of the triangles $PD'M$ and NME' implies that $x/b = a/y \Rightarrow yx = ab$. By the law of cosines,

$$|D'M| = \sqrt{y^2 + a^2 - 2ya\cos\alpha}, \qquad |E'M| = \sqrt{x^2 + b^2 - 2xb\cos\alpha},$$

It follows that the perimeter of the triangle $AD'E$ is

$$a + y + \sqrt{y^2 + a^2 - 2ya\cos\alpha} + b + x + \sqrt{x^2 + b^2 - 2xb\cos\alpha}.$$

$1°$ *Formalization.* In sum, we obtain the following formalization:

$$f_0(x, y) = x + \sqrt{x^2 + b^2 - 2xb\cos\alpha} + y$$

(p_{13})
$$+ \sqrt{y^2 + a^2 - 2ya\cos\alpha} \to \min,$$
$$f_1(x, y) = xy - ab = 0, \qquad x > 0.$$

If we express y in terms of x and substitute the result in the minimized function, then we arrive at the problem

(p'_{13}) $f_0(x) \to \min, \qquad x > 0,$

where $f_0(x) \to \infty$ as $x \to 0$ and $f_0(x) \to \infty$ as $x \to \infty$ (check this).

By the corollary to Weierstrass' theorem in the eleventh story, problem (p'_{13}), and therefore also (p_{13}), has a solution. We denote it by $(\hat{x}, \hat{y})$. We will use the Lagrange principle. The Lagrange function is

$$\mathscr{L}(x, y, \lambda_0, \lambda_1) = \lambda_0 f_0(x, y) + \lambda_1 f_1(x, y).$$

$2°$ *Necessary condition.*

$$\frac{\partial \mathscr{L}}{\partial x} = 0, \qquad \frac{\partial \mathscr{L}}{\partial y} = 0.$$

$3°$ *Finding the stationary points.*

$$\frac{\partial \mathscr{L}}{\partial x} = 0 \Rightarrow 1 - \frac{b \cos \alpha - x}{\sqrt{x^2 + b^2 - 2xb \cos \alpha}} + \lambda y = 0,$$

$$\frac{\partial \mathscr{L}}{\partial y} = 0 \Rightarrow 1 + \frac{y - a \cos \alpha}{\sqrt{y^2 + a^2 - 2ya \cos \alpha}} + \lambda x = 0.$$

$4°$ *Discussion.* We now explain the geometric significance of these relations. To this end we drop perpendiculars MR and MS to AC and AB, respectively. Then, as is clear from Figure 4.7 on page 35, we have

$$\frac{b \cos \alpha - x}{\sqrt{x^2 + b^2 - 2xb \cos \alpha}} = \frac{|RE|}{|ME|} = \cos \varphi,$$

$$\frac{y - a \cos \alpha}{\sqrt{y^2 + a^2 - 2ya \cos \alpha}} = \frac{|SD|}{|MD|} = \cos \psi.$$

If we multiply the first of the relations $3°$ by x and the second by y and make use of the relations in $4°$ and the equality $xy = ab$, then we obtain the relation $x(1 - \cos \varphi) = y(1 + \cos \psi)$. Applying the law of sines to the triangles MNE and DPM, we have

$$\frac{|ME|}{\sin \alpha} = \frac{x}{\sin \psi}, \qquad \frac{|MD|}{\sin \alpha} = \frac{y}{\sin \varphi}.$$

Hence

$$\frac{|ME|(1 - \cos \varphi)}{\sin \varphi} = \frac{|MD|(1 + \cos \psi)}{\sin \psi} \Rightarrow |ME| \tan \frac{\varphi}{2} = |MD| \tan \left(\frac{\pi - \psi}{2} \right).$$

The geometric significance of the last relation is the following: The perpendicular to DE at M and the bisectors of the exterior angles D and E intersect in one point O. In other words, the excircle to the triangle ADE passes through M. This is the answer we obtained in the fourth story.

14. Apollonius' problem. I said in the fourth story that problems on extrema are found in the works of all three of the greatest mathematicians

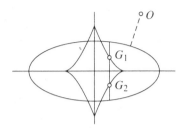

FIGURE 13.1

of antiquity—Euclid, Archimedes, and Apollonius. So far I have presented problems associated with Euclid and Archimedes.

I couldn't bring myself to present Apollonius' problem. Here is why.

The title of the greatest work of Apollonius (262?–190? B.C.) is *Conica*, or *Conics*. *Conica* is widely regarded as the apex of antique mathematics. Relevant to our topic is its fifth book. Here Apollonius "treats the shortest and the longest line segments from a point O to a conic. But he gives even more than he promises: he determines all the lines through O that intersect the conic at right angles (nowadays we call them normals), he investigates the positions of O for which there are two, three or four solutions... " By shifting the position of O "he determines the ordinates of the limiting points G_1 and G_2, at which the number of normals through O jumps from 2 to 4, or inversely." (See Figure 13.1.) The quoted lines are from B. L. van der Waerden's *Science awakening* (Noordhoff, 1954, pp. 260–261).

I did not want to formulate Apollonius' problem in the fourth story for two reasons. One reason is that the topic "conic sections" is not covered in high school. The other is that I could not imagine how to solve the problem discussed by van der Waerden while remaining within the framework of "old" elementary mathematics. And without presenting a solution, I didn't want to touch on this problem in Part One.

When talking about the brachistochrone in the seventh story I mentioned that the mathematicians of antiquity primarily considered lines, circles and conic sections (strictly speaking, these were also the curves investigated by Apollonius). But the curve separating the region in which the number of normals is two from the region where that number is four (it is called an *astroid*) belongs to an altogether different class of curves. It turned up first in the seventeenth century. It is difficult to see how it could be given without the use of the language of algebra. (Remember this when we find the equation of the astroid!)

We state the problems just mentioned as follows:

1. How does one determine the distance from a point to a conic section?
2. How many normals can one draw from a point to a conic?

We will solve these problems for an ellipse rather than for all conics.

The equation of an ellipse in a rectangular coordinate system is $(x_1/a_1)^2 + (x_2/a_2)^2 = 1$. We assume that $a_1 \geq a_2 > 0$, that is, the "width" of the ellipse is not less than its "height." If $a_1 = a_2$, that is if the width and height are equal, then the ellipse becomes a circle. We will now solve the first problem.

$1°$ *Formalization.* Let O be a point with coordinates (ξ_1, ξ_2). The distance from a point with coordinates (ξ_1, ξ_2) to one with coordinates (x_1, x_2) is $((x_1 - \xi_1)^2 + (x_2 - \xi_2)^2)^{1/2}$. It is convenient to minimize the square of the distance rather than the distance itself. In sum, we obtain the problem

$$f_0(x_1, x_2) = (x_1 - \xi_1)^2 + (x_2 - \xi_2)^2 \to \min,$$
$$f_1(x_1, x_2) = (x_1/a_1)^2 + (x_2/a_2)^2 - 1 = 0.$$

The functions f_0 and f_1 and their partial derivatives are continuous. Since $-a_1 \leq x_j \leq a_1$, $j = 1, 2$, the set of admissible points is bounded. Hence a solution $\hat{x} = (\hat{x}_1, \hat{x}_2)$ exists, and we can use the Lagrange principle. The Lagrange function is

$$\mathscr{L} = \lambda_0 f_0 + \lambda_1 f_1.$$

$2°$ *Necessary condition.*

$$\frac{\partial \mathscr{L}}{\partial x_j} = 0, \qquad j = 1, 2.$$

$3°$ *Finding the stationary points.*

$$\frac{\partial \mathscr{L}}{\partial x_1} = 0 \Rightarrow \lambda_0(\hat{x}_1 - \xi_1) + \lambda_1 \hat{x}_1/a_1^2 = 0,$$
$$\frac{\partial \mathscr{L}}{\partial x_2} = 0 \Rightarrow \lambda_0(\hat{x}_2 - \xi_2) + \lambda_1 \hat{x}_2/a_2^2 = 0.$$

If we suppose that $\lambda_0 = 0$, then $\lambda_1 \neq 0$ (the Lagrange multipliers cannot all be zero). But then our equations imply that $\hat{x}_1 = \hat{x}_2 = 0$, that is $0 = f_1(\hat{x}_1, \hat{x}_2) = f_1(0, 0) = -1$. Hence $\lambda_0 \neq 0$ and we can put $\lambda_0 = 1$. We set $\lambda_1 = \lambda$. From our equations it follows that

$$(\hat{x}_j - \xi_j) + \lambda \hat{x}_j/a_j^2 = 0, \qquad j = 1, 2,$$
$$\Rightarrow \hat{x}_j = \frac{\xi_j a_j^2}{(a_j^2 + \lambda)}, \qquad j = 1, 2.$$

Substituting these relations in the equation of the ellipse we obtain the equation

$$\varphi(\lambda) = \frac{\xi_1^2 a_1^2}{(a_1^2 + \lambda)^2} + \frac{\xi_2^2 a_2^2}{(a_2^2 + \lambda)^2} = 1.$$

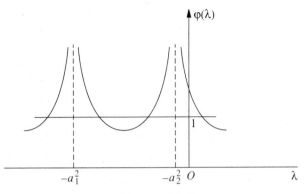

FIGURE 13.2

4° *Discussion.* The number of stationary points in the problem (that is, points corresponding to the values of λ that satisfy the equation $\varphi(\lambda) = 1$) does not exceed four (because we have obtained an equation of degree four; see also Figure 13.2). In Figure 13.2, we have the case $\varphi(0) > 1$. Since $\varphi(0) = \xi_1^2/a_1^2 + \xi_2^2/a_2^2$, the point (ξ_1, ξ_2) lies outside the ellipse.

It is clear that it is impossible to write down the solutions of this equation in some simple, explicit form. But now that we have at our disposal many computing tools, we can find solutions of the equation $\varphi(\lambda) = 1$ very quickly and with an arbitrary degree of precision. Once the roots λ_i of the equation have been computed, it will be necessary to find the corresponding points $(x_1(\lambda_i), x_2(\lambda_i))$, substitute these values in f_0, and find the smallest of the resulting numbers.

The first of our two problems has been solved. The geometric significance of the relations $(\hat{x}_j - \xi_j) + \lambda \hat{x}_j / a_j^2$ is that the vector $\xi - \hat{x}$, joining O and a minimal point of the ellipse, is proportional to the gradient of f_1 at $\hat{x}$, that is, the vector $\xi - \hat{x}$ lies on the normal to the ellipse. This fact was first established by Apollonius.

Now we'll tackle the second problem. We will derive the equation of the "dividing" curve that separates the region where one can lead two normals through a point from the region where one can lead four. From Figure 13.2 it is easy to see that this division occurs for values of λ for which $\varphi(\lambda) = 1$ and $\varphi'(\lambda) = 0$, because that is when the curve $y = \varphi(\lambda)$ touches the line $y = 1$. In other words, we must eliminate λ from the relations

$$\varphi(\lambda) = \frac{\xi_1^2 a_1^2}{(a_1^2 + \lambda)^2} + \frac{\xi_2^2 a_2^2}{(a_2^2 + \lambda)^2} = 1,$$

$$\varphi'(\lambda) = -\frac{\xi_1^2 a_1^2}{(a_1^2 + \lambda)^3} - \frac{\xi_2^2 a_2^2}{(a_2^2 + \lambda)^3} = 0.$$

From the second of these relations we obtain

$$a_1^2 + \lambda = A(\xi_1 a_1)^{2/3}, \qquad a_2^2 + \lambda = -A(\xi_2 a_2)^{2/3},$$

where

$$A = (a_1^2 - a_2^2)/[(\xi_1 a_1)^{2/3}(\xi_2 a_2)^{2/3}].$$

Substituting these relations in the equation $\varphi(\lambda) = 1$, we arrive at the equation of the dividing curve

$$(\xi_1 a_1)^{2/3} + (\xi_2 a_2)^{2/3} = (a_1^2 - a_2^2)^{2/3}.$$

This is the equation of the astroid discussed earlier (how could Apollonius have obtained it?). Outside the astroid each point has two normals, inside it, four (in particular, obviously, at the center of the ellipse), and on the astroid itself, three (except at the vertices, where there are two normals).

Finally we have arrived at a result that was first obtained in the second century B.C.

In 1975 the first all-Soviet Olympiad, "The student and scientific-technical progress," was organized. The Olympiad was a gathering of about 100 of the best mathematics majors from the Soviet republics. One of the problems set before the participants was Apollonius' problem: How many normals can one lead from a point to an ellipse? Just one person could cope with this problem. To tell the truth, the organizers thought that after twenty-two centuries more would have been achieved.

In the summer of 1984, I coached high school students preparing to take part in the international mathematical Olympiad. The topic of the session was "Mathematical analysis." To demonstrate the power of mathematical analysis, I decided to tell the students about topics that you have encountered in Part Two of this book. During the study sessions we arranged a kind of contest between analysis and geometry.I would suggest a problem, the students would solve it geometrically, and I would solve it analytically. I was convinced of the superiority of analysis and hoped for an easy victory. But matters turned out to be anything but simple. My listeners were true lovers of geometry and remarkably well-trained problem-solvers. It was a small matter for these youngsters to think of unexpected and very elegant solutions that were—I thought—anything but easy to find. And they regarded them as trivial. There was no easy victory. Nor could it be said that it was a fiasco for mathematical analysis. I will now present three problems from my coaching with the high school students in which, I think, the "theory" acquitted itself very well indeed.

The solutions I present rely on analysis. Readers are urged to try and find "purely geometric" solutions that are obviously simpler than these.

The first problem was given at the all-Soviet mathematical Olympiad for high school students in 1980. Its author is I. F. Šarygin, who is known for his

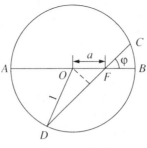

FIGURE 13.3

remarkable ability to invent beautiful geometric problems (see [10R], Problem 349).

15. *Given a unit circle. Through a given point F on a diameter AB, pass a chord CD so that the quadrilateral $ACBD$ has maximal area.*

$1°$ *Formalization.* Let O be the center of the circle, put $|OF| = a$, and denote the angle CFB by φ. (See Figure 13.3.) Recall that the area of a cyclic quadrilateral is half the product of its diagonals by the sine of the angle between them. It is clear that $|CD| = 2\sqrt{1 - a^2 \sin^2 \varphi}$. This leads to the formalization

$$\sqrt{1 - a^2 \sin^2 \varphi} \, \sin \varphi \to \max, \qquad 0 \le \varphi \le \pi/2.$$

By making the substitution $a \sin \varphi = \sqrt{z}$, we obtain the problem

$$f(z) = (1 - z)z \to \max, \qquad 0 \le z \le a^2.$$

Weierstrass' theorem implies the existence of a solution.

$2°$ *Necessary condition.* Fermat's theorem: $f'(\hat{z}) = 0$.

$3°$ *Finding the critical points.* There is just one stationary point: $\hat{z} = 1/2$ (if $a^2 > 1/2$). The critical points are: $\{0, a^2\}$ if $a^2 \le 1/2$ and $\{0, \frac{1}{2}, a^2\}$ if $a^2 > 1/2$.

$4°$ *Discussion.* By looking at the values of f at the critical points, we arrive at the answer: if $0 \le a \le 1/\sqrt{2}$, then $\hat{z} = a^2$, that is, $\hat{\varphi} = \pi/2$; if $1/\sqrt{2} < a \le 1$, then $\hat{z} = 1/2$, that is, $\hat{\varphi} = \arcsin(1/\sqrt{2})$.

It was the students who gave me this problem to solve. They were convinced that their geometric solution would be incomparably simpler than the analytic one. But could any solution be simpler than ours?

The next problem is also due to I. F. Šarygin. He invented it especially for the study session with this group. (The problem was to have been used in preliminary competitions, but this did not happen, so that the problem was

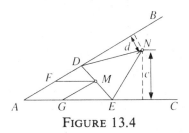

FIGURE 13.4

unknown to my listeners. Knowing I. F. Šarygin, who coached them, they looked forward to the usual easy triumph of geometry.) This is the problem in question (see [10R], Problem 348).

16. Given an angle BAC and two points M and N in its interior, pass a segment DE through M (using ruler and compass) such that the area of the quadrilateral $ADNE$ is minimal. (See Figure 13.4.)

$1°$ *Formalization.* Pass segments MF and MG parallel to the sides AC and AB, respectively, and denote their lengths by a and b. Drop perpendiculars from N to AB and AC and denote their lengths by d and c. Then twice the area of $ADNE$ is equal to $(b+x)d+(a+y)c$, where $x=|FD|$ and $y=|GE|$. Also, $xy=ab$; this follows directly from the similarity of the triangles DFM and MGE. Hence

$$f_0(x,y)=(b+x)d+(a+y)c \to \min,$$
$$f_1(x,y)=xy-ab=0.$$

The existence of a solution $(\hat{x},\hat{y})$ follows (think about it!) from Weierstrass' theorem (f_0 goes to infinity with x). The functions f_0 and f_1 and their partial derivatives are continuous, so that we can use the Lagrange principle. The Lagrange function is

$$\mathscr{L}=\lambda_0 f_0+\lambda_1 f_1.$$

$2°$ *Necessary condition.*

$$\frac{\partial \mathscr{L}}{\partial x}=0, \qquad \frac{\partial \mathscr{L}}{\partial y}=0.$$

$3°$ *Finding the stationary points.*

$$\frac{\partial \mathscr{L}}{\partial x}=0 \Rightarrow \lambda_0 d+\lambda_1 \hat{y}=0,$$
$$\frac{\partial \mathscr{L}}{\partial y}=0 \Rightarrow \lambda_0 c+\lambda_1 \hat{x}=0.$$

Clearly $\lambda_0 \neq 0$ (otherwise $\hat{x}=\hat{y}=0$, that is, $\hat{x}\hat{y}=0 \neq ab$), so that we can put $\lambda_0=1$. In sum, we obtain the system of equations

$$\hat{x}\hat{y}=ab, \qquad \hat{x}/\hat{y}=c/d.$$

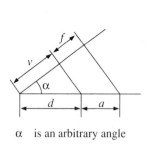

α is an arbitrary angle

FIGURE 13.5   FIGURE 13.6

$4°$ *Construction.* Using the relation $f/b = a/d$, we construct (using ruler and compass) the segment f (a glance at Figure 13.5 will remind you how to do this). Using the relation $\hat{x}^2 = cf \Leftrightarrow \hat{x} = \sqrt{cf}$, we construct $\hat{x}$ (again by means of ruler and compass—see Figure 5.2 on p. 40).

There are purely geometric solutions. In particular there is the very beautiful solution due to... the author! These solutions lead to other construction methods. Find them and compare. My students have not managed to come up with a simpler geometric solution!

17. *Among all pyramids with given base and height find the one with least lateral area.*

$1°$ *Formalization.* (See Figure 13.6.) Let the base of the pyramid be a triangle $A_1 A_2 A_3$ with sides of length a_1, a_2, and a_3 $(a_j > 0, \ j = 1, 2, 3)$, and let O be the projection of the vertex on the base plane. Denote by H the height of the pyramid and by h_1, h_2 and h_3 the distances from O to the lines containing the sides $A_2 A_3$, $A_1 A_3$, and $A_1 A_2$, respectively (they are to be taken with a plus sign if O lies in the same halfplane as the triangle $A_1 A_2 A_3$ and with a minus sign otherwise). Then we have the familiar (and obvious) equality $a_1 h_1 + a_2 h_2 + a_3 h_3 = 2S$, where S is the base area. Also, the lateral area of the pyramid is $((a_1 \sqrt{H^2 + h_1^2} + a_2 \sqrt{H^2 + h_2^2} + a_3 \sqrt{H^2 + h_3^2})/2) + S$. This leads to the problem

$$f_0(h_1, h_2, h_3) = a_1 \sqrt{H^2 + h_1^2} + a_2 \sqrt{H^2 + h_2^2} + a_3 \sqrt{H^2 + h_3^2} \rightarrow \min,$$
$$f_1(h_1, h_2, h_3) = a_1 h_1 + a_2 h_2 + a_3 h_3 - 2S = 0.$$

By Weierstrass' theorem (think this through!) the problem has a solution (f_0 is increasing at infinity). The functions f_0 and f_1 and their partial derivatives are continuous. This means that a solution $\hat{h} = (\hat{h}_1, \hat{h}_2, \hat{h}_3)$ exists, and we can use Lagrange's principle. The Lagrange function is $\mathscr{L} = \lambda_0 f_0 + \lambda_1 f_1$.

2° *Necessary condition.*

$$\frac{\partial \mathscr{L}}{\partial h_j} = 0, \qquad j = 1, 2, 3.$$

3° *Finding the stationary points.*

$$\frac{\partial \mathscr{L}}{\partial h_j} = 0 \Rightarrow \lambda_0 \frac{a_j \hat{h}_j}{\sqrt{H^2 + \hat{h}_j^2}} + \lambda_1 a_j = 0, \qquad j = 1, 2, 3.$$

Clearly, $\lambda_0 \neq 0$, and we can assume that $\lambda_0 = 1$.

4° *Discussion.* Dividing the resulting equations by a_j, we immediately find that $\hat{h}_1 = \hat{h}_2 = \hat{h}_3$, that is, *the projection of the altitude is the center of the incircle.*

After a little reflection, one of my listeners gave the right answer. I asked him to come up to the board and expected to hear once more a "purely geometric" solution accompanied by ironic comments. Quite unexpectedly, I saw the functions f_0 and f_1, the Lagrange function, its partial derivatives and—the answer.

Long live mathematical analysis! Don't you agree?

I waited impatiently for news from the twenty-fifth international mathematical Olympiad that took place in Prague in 1984. First I learned that our team performed with distinction; the youngsters won five first prizes and one second prize and collected 225 points. No team ever performed with such distinction in all the history of the Olympiad. Then I saw the problems from the competition. Of course, I was especially interested in the problems in which one could use the methods of investigation of extremal problems. One such problem appeared.

18. Let x, y, and z be nonnegative real numbers with $x + y + z = 1$. Show that $0 \leq xy + yz + xz - 2xyz \leq 7/27$.

1° *Formalization.*

$$f_0(x, y, z) = xy + yz + xz - 2xyz \to \max(\min),$$
$$f_1(x, y, z) = x + y + z - 1 = 0, \qquad x \geq 0, \qquad y \geq 0, \qquad z \geq 0.$$

By Weierstrass' theorem, the problem allows a solution in the case of a maximum as well as a minimum. Suppose that $(\hat{x}, \hat{y}, \hat{z})$ is a solution with nonzero entries. Then this solution will yield a local extremum for the problem $f_0(x, y, z) \to \max(\min)$, $f_1(x, y, z) = 0$, and one can apply to it the Lagrange multiplier rule.

$2°$ *Necessary condition (for the Lagrange function $\mathscr{L} = \lambda_0 f_0 - \lambda_1 f_1$).*

(i)
$$\frac{\partial \mathscr{L}}{\partial x} = 0 \Leftrightarrow \lambda_0(y + z - 2yz) = \lambda_1,$$

(ii)
$$\frac{\partial \mathscr{L}}{\partial y} = 0 \Leftrightarrow \lambda_0(x + z - 2xz) = \lambda_1,$$

(iii)
$$\frac{\partial \mathscr{L}}{\partial z} = 0 \Leftrightarrow \lambda_0(x + y) - 2xy) = \lambda_1.$$

Clearly, $\lambda_0 \neq 0$ (otherwise $\lambda_1 = 0$). We put $\lambda_0 = 1$.

$3°$ *Finding the stationary points.* Subtracting the second equation from the first (with $\lambda_0 = 1$), we get

$$y - x - 2z(y - x) = 0 \Rightarrow z = \frac{1}{2}, \quad \text{or } y = x.$$

Similarly, $y = z$, or $x = 1/2$, and $x = z$, or $y = 1/2$.

$4°$ *Discussion.* If one of the numbers, say z, is $1/2$, then $f_0(x, y, 1/2) < 7/27$. If none of these numbers is $1/2$, then there is a unique stationary point $\hat{x} = \hat{y} = \hat{z} = 1/3$ and $f_0(1/3, 1/3, 1/3) = 7/27$. Finally, if the solution has a zero component, say $\hat{x} = 0$, then $0 \leq f_0(x, y, z) = yz \leq 1/4$. Answer: the maximum $7/27$ is attained for $\hat{x} = \hat{y} = \hat{z} = 1/3$, and the minimum, zero, is attained for, say, $\hat{x} = \hat{y} = 0$, $\hat{z} = 1$.

One more geometric problem.* It is tied to a memory. This was a long time ago, in fact, more than 30 years ago. The leader of the mathematical circle smiled enigmatically and asked: "One tetrahedron lies inside another. Can the sum of its edges be greater than the sum of the edges of the outer tetrahedron?" At first, this seemed a total impossibility. How can anything about the inner tetrahedron be greater? But it turns out that the sum of the edges can indeed be larger for the smaller tetrahedron. During the second round of the sixteenth all-Soviet mathematical Olympiad, tenth-grade students were given the following problem.

19. *The vertices of a tetrahedron $KLMN$ lie inside, on the faces, or on the edges of another tetrahedron $ABCD$. Show that the sum of the lengths of all edges of the tetrahedron $KLMN$ is less than $4/3$ of the sum of the lengths of all edges of the tetrahedron $ABCD$.*

This is an interesting example. It shows that it is sometimes possible to solve a problem without the standard investigation, for a great deal becomes immediately clear after the formalization stage.

The tetrahedron $ABCD$ is a convex, closed, and bounded set. Denote it

---

* For more on this problem see the author's paper in the journal, *Kvant*, 1983, 1, pp. 22–25 (in Russian).

by X. X can be given by means of four inequalities:

$$X = \{x = (x_1, x_2, x_3) \mid \langle x, a^i \rangle \leq \alpha_i, \ i = 1, 2, 3, 4\}$$

where $\langle x, y \rangle$ is the scalar product of x and y.

Our problem can be formalized as follows:

$$f(x^1, x^2, x^3, x^4)$$
$$= |x^1 - x^2| + |x^1 - x^3| + |x^1 - x^4| + |x^2 - x^3| + |x^2 - x^4| + |x^3 - x^4|$$
$$\to \max, \qquad x^i \in X, \qquad i = 1, 2, 3, 4.$$

Here $x^k = (x_1^k, x_2^k, x_3^k)$, $k = 1, 2, 3, 4$, are points in three-dimensional space, and $|x - y|$ is the distance, in that space, between x and y.

The function f is a continuous function of $4 \cdot 3 = 12$ variables. The tetrahedron is a closed and bounded set given by four inequalities. By Weierstrass' theorem, the problem has a solution. Denote it by $(\hat{x}^1, \hat{x}^2, \hat{x}^3, \hat{x}^4)$. In view of the strict convexity of f (that follows—think this through—from the properties of the distance function between a fixed point and a given point), it follows immediately that the points x^i must coincide with the vertices of X. In fact, if, for example, $\hat{x}^1$ is not a vertex, then there is a segment $[y, z]$ where y and z are points in X and $\hat{x}^1 = (y + z)/2$. But then, in view of the property of a strictly convex function,

$$f(\hat{x}^1, \hat{x}^2, \hat{x}^3, \hat{x}^4) = f\left(\frac{y+z}{2}, \hat{x}^2, \hat{x}^3, \hat{x}^4\right)$$
$$< \tfrac{1}{2}[f(y, \hat{x}^2, \hat{x}^3, \hat{x}^4) + f(z, \hat{x}^2, \hat{x}^3, \hat{x}^4)].$$

This implies that at one of the points $(y, \hat{x}^2, \hat{x}^3, \hat{x}^4)$ and $(z, \hat{x}^2, \hat{x}^3, \hat{x}^4)$ the function takes on a value larger than $f(\hat{x}^1, \hat{x}^2, \hat{x}^3, \hat{x}^4)$. This is obviously a contradiction.

All that remains now is a simple sorting step. We denote the perimeters of the tetrahedra $KLMN$ and $ABCD$ by P_{KLMN} and P_{ABCD}, respectively.

If all points K, L, M, and N are different, then the tetrahedron $KLMN$ coincides with $ABCD$ and all is clear, for

$$P_{KLMN} = P_{ABCD} < \tfrac{4}{3} P_{ABCD}.$$

Suppose that no more than two of the vertices of the tetrahedron $KLMN$ coincide, say, $K = L = A$. Then there are two possibilities.

1. The two other vertices also coincide, say $M = N = B$, in which case the triangle inequality implies that

$$P_{KLMN} = 4|AB| = \tfrac{4}{3} \cdot 3|AB|$$
$$< \tfrac{4}{3}[|AB| + (|AC| + |CB|) + (|AD| + |DB|)] < \tfrac{4}{3} P_{ABCD}.$$

2. The vertices M and N are different, say, $M = B$, $N = C$; then the triangle inequality implies that

$$|AB| < |AD| + |DB|, \qquad |AC| < |AD| + |DC|;$$
$$|AD| < \tfrac{1}{3}[|AD| + (|AC| + |CD|) + (|AB| + |BD|)]$$
$$< \tfrac{1}{3}P_{ABCD} \Rightarrow P_{KLMN}$$
$$= 2|AB| + 2|AC| + |BC|$$
$$< |AB| + |AD| + |DB| + |AC| + |AD| + |DC| + |BC|$$
$$= P_{ABCD} + |AD| < \tfrac{4}{3}P_{ABCD}.$$

If three vertices coincide, say, $K = L = M = A$, $N = B$, then the inequality in **1** above implies that $P_{KLMN} = 3|AB| < 4|AB| < 4/3 P_{ABCD}$. Thus, the problem has been solved.

It is easy to see that the number $4/3$ in the statement of the problem cannot be decreased. It is attained for the degenerate tetrahedron $ABCD$ whose three vertices A, B, and C coincide. Indeed, suppose that A, B, and C coincide with A. Then in the tetrahedron $KLMN$ two vertices coincide with A and two with D. But then $P_{ABCD} = 3|AD|$ and $P_{KLMN} = 4|AD|$.

14

What Happened Later
in the Theory of Extremal Problems?

We shall consider the simplest maximum and minimum problem that points to a natural transition from functions of a finite number of variables to magnitudes that depend on an infinite number of variables.

V. Volterra

The methods I set forth require neither constructions nor geometric or mechanical considerations. They require only algebraic operations subject to a systematic and uniform course.

J. Lagrange

An old French mathematician said: "A mathematical theory can be regarded as perfect only if you are prepared to present its contents to the first man in the street."

D. Hilbert

1. On the history of mathematical analysis. The development of methods of solution of maximum and minimum problems is inextricably linked to the history of mathematical analysis. We have touched this topic many times.

Now we will tie together much of what was told before.

Recall that at first maximum and minimum problems were solved individually, each problem giving rise to a particular solution. At the beginning of the seventeenth century, there arose the need to find some general methods of investigation of extremal problems. Descartes attempted to find algebraic means of locating maxima and minima. Fermat was the first to employ for such purposes what we now call the differential calculus. According to his own words, he had discovered his method as early as 1629. However, the first relatively detailed account of the method is found in his letters to Roberval (sent in 1636) and to Mersenne, who forwarded his copy to Descartes. Descartes received it in 1638.

"The whole theory of evaluation of maxima and minima presupposes ... the following single rule," wrote Fermat, who then went on to present the essence of his method as discussed in the tenth story.

The reader would do well to read Fermat's paper (see pp. 223–227 of *A Source Book in Mathematics*, edited by D. J. Struik, Harvard University Press, 1969) in order to find out how he managed to describe his method without using the yet-to-be-invented notion of a derivative.

Fermat supported his theoretical arguments with an example: "*To divide the segment AC at E so that the rectangle with sides AE and AC may be maximal (in terms of area).*" It is easy to see that this is the very same problem that Euclid posed and solved geometrically in his *Elements* (see the fourth story). This example (in Fermat's formulation) was analyzed in the fifth story.

In 1671 Newton completed his *Of the methods of series and fluxions with application to the geometry of curves* [10]. This work was not published until 1736. Here Newton laid the foundations of the differential and integral calculus and of the theory of infinite series. Of course, Newton also paid attention to finding maxima and minima. He mentions Fermat's method in passing without mentioning Fermat's name. He writes: "... seek its fluxion [that is, the derivative of a quantity] and set it equal to nothing" [10]. Newton solves two significant examples involving implicit functions, one of which contains a radical. Then he writes: "Using the method of solution of this problem one can obtain the solutions of the following problems," and lists nine geometric problems that he can solve. And again the first of these is a problem equivalent to Euclid's.

In 1684 Leibniz published a work in which he also laid the foundations of mathematical analysis. Its very title, beginning with the words *A new method for maxima and minima* ... , shows the importance of the role of the problem of finding extrema in the formation of modern mathematics. In his paper Leibniz not only finds the necessary condition $f'(x) = 0$, but he also uses the second differential to distinguish between a maximum and a minimum (incidentally, this was also known at the time to Newton). With the help of the relation $f'(x) = 0$, Leibniz solves a number of concrete problems, including the derivation of Snel's law (third story).

Leibniz' works were significantly ahead of their time. In them one can already discern the thought of linear approximation of functions, of the connection between the tangent and the derivative. This idea underwent an interesting evolution, described in §5.

The research of Fermat, Newton, and Leibniz promoted the emergence of the method of finding extrema of functions of one variable. It seems that it would have been natural to study next extrema of functions of two variables, of three variables, and so on. But this is not what happened. The history of analysis made a kind of zigzag and immediately embarked on the study of functions of infinitely many variables. It took decades for it to return to strictly finite-dimensional problems.

In Newton's problem (eighth story), in the brachistochrone problem (seventh story), and in the classical isoperimetric problem (second story), "arbitrary curves" are tested. These curves cannot be given by one, two, or any arbitrary finite number of parameters. Their "arbitrary rule" includes "an infinitely large number of variables." Small wonder we were unable to solve these three problems in the previous story.

The elaboration of a theory of problems similar to these three began at the end of the seventeenth century. A special "calculus" of such problems was created, taking shape in the eighteenth and nineteenth centuries in the works of Euler, Lagrange, Weierstrass, and others. This theory came to be known as *the calculus of variations*.

Analysis of functions of a finite number of variables was developed somewhat later. And then, relatively recently, it was understood that mathematical analysis of an infinite number of variables is not, in principle, more complex than finite-dimensional analysis. Once again the thought of creating an infinite-dimensional analysis was launched by the need to solve extremal problems (see Volterra's epigraph; Volterra has in mind the classical isoperimetric problem). Let's examine this topic.

2. What is a function of an infinite number of variables? In the ninth story we discussed the question: What is a function? First we looked at functions of one variable, where we associate to a single number x a number y in accordance with a definite rule. Then we examined the matter of a function of two variables, where we associate to a pair of numbers (x_1, x_2) a number y (again, in accordance with a definite rule). Finally we also discussed functions of n variables. But even before the ninth story we encountered (on a number of occasions) functions of infinitely many variables, where the variables were themselves functions. (Recall formula (2) in the seventh story and formula $(8')$ in the eighth.) Mathematicians had studied such *functionals* (the usual name of functions defined on functions) for almost two hundred years before they learned to handle functions of an infinite number of variables with as much dispatch as functions of one variable.

Consider the set (in mathematics we also say the *space*) of all functions continuous on a segment $[a, b]$ of the real line. This space is denoted by $C([a, b])$. Now let these functions play the role of the variable. We'll try to interpret this fact. We recall the definition of a function of one variable, that is, a function defined on the real line $\mathbb{R}$ (say the function $y = y(x) = \sqrt{1 + x^2}$). This, we remember, is a rule that enables us to obtain the number y for a given number x (in the concrete example we must square x, add one, and take the square root of the sum).

Now we'll try to understand the nature of a function $F(y)$ on the space $C([a, b])$. By the very meaning of "function" this must be a rule that enables us to compute the number $F(y)$ for a given continuous function $y(x)$ on $[a, b]$. Let's consider some examples. We will set $a = 0$ and $b = 1$ for definiteness.

EXAMPLE 1. $F_1(y) = 2y(0)$.

What has been prescribed? Given a function $y(x)$ we must first compute its value at zero and then multiply this value by 2. Let's recall some familiar functions and compute the values $F_1(y)$ for them. Thus if $y(x) = x$, then $y(0) = 0$ and therefore $F_1(y) = 0$; if $y(x) = \sqrt{1 + x^2}$ then $y(0) = 1$ and $F_1(y) = 2$; if $y(x) = 5 \cdot 2^x$, then $y(0) = 5$ and $F_1(y) = 10$.

Now imagine the following game: I give you, one after another, a number of functions, and for each you are to compute $F_1(y)$. For example, I give you $y(x) = 3\cos(x + 2)$, or $5\ln(x + 3)$, or some such. I think that in these cases you'll find it easy to compute a value for $F_1(y)$. To understand the rules of this game is to understand what is meant by the function F_1.

EXAMPLE 2. $F_2(y) = \int_0^1 y(x)\,dx$. (This function is just *the area under the graph* of $f(x)$.) Again, let's play the same game. I give you $y(x) = 1$ and you compute for me $F_2(y) = 1$; my move is $y(x) = x$ and your response is $F_2(y) = 1/2$; my move is $y(x) = \sin x$ and your response is $F_2(y) = 1 - \cos 1$, and so on. We have become acquainted with a very important functional—*the area functional*.

We can think of an even cleverer example.

EXAMPLE 3. $F_3(y) = (2y(0))^3 - (\int_0^1 y(x)\,dx)^2$. Here the prescription is more involved. Given a function $y(x)$ you must compute $y(0)$, multiply this number by 2, cube the resulting number, then compute the integral of $y(x)$ on $[0, 1]$, square it, and subtract this "square" from the earlier "cube." For example, suppose you are given $y(x) = x$. Then your response is $-1/4$. If you are given $y(x) = 5 \cdot 2^x$, then getting the right response will be a bit difficult. What counts, however, is that "in principle" you can carry out the task and obtain the number $F_3(y)$ for each given (well-defined) function $y(x)$.

In the ninth story I wrote, "Now let's define and represent some of the most important functions ... ." Now that we face an infinite-dimensional space, any "representing" is difficult. But "defining" is a different matter, and we'll give it a try.

The simplest function is a constant, $F(y) \equiv c$. This function associates to every continuous function $y(x)$ one and the same number c.

Next in the order of complexity are linear functions. What does "linear function" of a function mean? It means that it associates to the sum of any functions a sum of numbers (that is, $F(y_1 + y_2) = F(y_1) + F(y_2)$) and, in addition, $F(ay) = a F(y)$ for any number a and any function $y(x)$.

The functions in the Examples 1 and 2 above are linear. The function in the third example is not linear. Here, in the infinite-dimensional case, there is an abundance of linear functions. (In a sense, there are "more" linear functions than continuous functions. Thus, if φ is a continuous function, then we can associate to it the linear functional

$$F_\varphi(y) = \int_0^1 \varphi(x) y(x) \, dx \, ,$$

whereas the linear functional $F_1(y) = 2y(0)$ cannot be so represented.)

Infinite-dimensional analysis studies functions of "an infinite number of variables," more precisely, functionals on infinite-dimensional space (like the space $C([a, b])$. We'll give an example of another important infinite-dimensional space with which the calculus of variations operated, basically, for two centuries. This is the space $C^1([a, b])$ of continuously differentiable functions $y(x)$, that is, functions $y(x)$ that are continuous, together with their derivatives, on the segment $[a, b]$.

On the space $C^1([a, b])$ there are defined functionals that have important geometric or physical meanings. Let's look at some examples.

EXAMPLE 4. The "length" functional:

$$L(y) = \int_a^b \sqrt{1 + (y'(x))^2} \, dx.$$

EXAMPLE 5. The functional of Johann Bernoulli—"the time of motion along a curve" (see formula (2) in the seventh story):

$$T(y) = \int_0^a \frac{\sqrt{1 + (y'(x))^2}}{\sqrt{2gy(x)}} \, dx.$$

EXAMPLE 6. Newton's functional—resistance to motion in a rare medium (see formula $(8')$ in the eighth story):

$$F(y) = 2K \int_0^R \frac{x \, dx}{1 + (y'(x))^2}.$$

There is an endless supply of such examples. Rather than give more of them, let's address the question of *how one poses extremal questions for functionals*. In the ninth story we saw that to formulate precisely an extremal

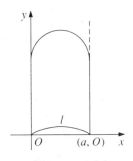

FIGURE 14.1

problem we must describe the function to be maximized or minimized as well as the constraints. Recall that the constraints are usually given by equalities and inequalities. In the infinite-dimensional case nothing changes. Here we must also describe the functional to be maximized or minimized (and thus also the space on which it is defined) and the constraints.

We'll now formalize those problems from Part One that we have so far not solved in Part Two. In all cases we'll consider problems in the space C^1.

Dido's problem. Recall the story of Dido (see the second story). After an analysis of the situation of the Phoenician princess, we will submit the following two possibilities for stating the optimization problem.

A) DIDO'S FIRST PROBLEM, OR THE CLASSICAL ISOPERIMETRIC PROBLEM. *To determine the optimal shape of a piece of land that would, for a given length of its perimeter l, have maximal area.*

We considered this problem in the second story. Other formulations can be obtained if we make the reasonable assumption that Dido wished to secure access to the sea. For the sake of simplicity, let's consider the case of a rectilinear shoreline and assume that Dido was shown boundaries she was not to cross. (See Figure 14.1) Then we obtain

B) DIDO'S SECOND PROBLEM. *Among all arcs of length l in the halfstrip $0 \le x \le a$, $y \ge 0$ with prescribed endpoints $(0,0)$ and $(0,a)$, find an arc that, together with the segment $y = 0$, $0 \le x \le a$, bounds a figure of maximal area.*

We'll limit ourselves to the formalization of the second problem.[*] Let $y = y(x)$ be the equation of the arc. Earlier we encountered the functionals "area" and "length." Bearing in mind their definitions, we arrive at the following

---

[*] The formalization of the first problem requires consideration of functionals of a pair of functions. We'll describe such problems at the end of this story. Also, the solution of the first problem is easy to obtain from that of the second.

formulation:

$$S(y) = \int_0^a y(x)\,dx \to \max, \qquad L(y) = \int_0^a \sqrt{1 + (y'(x))^2}\,dx = l$$

with boundary conditions $y(0) = 0$, $y(a) = 0$.

The functional to be maximized is area. The constraint is given by the equality $L(y) = l$, where $L(y)$ is the length functional. The boundary conditions are also given by equalities $\Gamma_1(y) = 0$, $\Gamma_2(y) = 0$, where $\Gamma_1(y) = y(0)$ and $\Gamma_2(y) = y(a)$.

The brachistochrone problem. Essentially, we formalized this problem in the seventh story. Recalling formula (2) from that story, we obtain the required formalization

$$T(y) = \int_0^a \frac{\sqrt{1 + (y'(x))^2}}{\sqrt{2gy(x)}}\,dx \to \min$$

with the boundary conditions $y(0) = 0$ and $y(a) = b$.

Here the functional to be minimized is the Bernoulli functional.

Newton's problem. Actually, we formalized this problem in the eighth story. The formalization is

$$F(y) = 2K \int_0^R \frac{x\,dx}{1 + (y'(x))^2} \to \min, \qquad y'(x) \geq 0,$$

with boundary conditions $y(0) = 0$ and $y(R) = H$.

We pay special attention to the constraint $y'(x) \geq 0$—the monotonicity condition. We have encountered it only once before.

Recall that we solved all these problems in different ways. But all solutions had one thing in common: In all of them, we approximated the curve by a polygonal line and in this way reduced the problem to a finite-dimensional one.

Our method of solution of problems of this kind was implemented by Euler, whose predecessor was Leibniz. This method and its modifications are known as the *direct methods of the calculus of variations.* They are used to this day for the numerical solution of problems in the calculus of variations.

3. Problems of the calculus of variations and Lagrange's principle for them. We have used the term "calculus of variations" many times, and now it is time to make it precise. Suppose we are given some function $f(x, y, z)$, a continuous function of three variables. We consider the functional

$$F(y) = \int_a^b f(x, y(x), y'(x))\,dx.$$

This functional can be considered in different spaces, but most often it is investigated in the space C^1. *By the calculus of variations we mean the section of the theory of extremal problems devoted to the study of maxima and minima of such functionals for various constraints.* (I'll say more about this later.)

Let us see what we must do to obtain the number $F(y)$ for a given function $y(x)$ in $C^1([a, b])$. First we must differentiate $y(x)$. Then we must put $y(x)$ for the second argument and $y'(x)$ for the third. The result is a function of one variable that associates to a number x the number $f(x, y(x), y'(x))$. Finally, we must integrate this function. In this way we obtain the required number $F(y)$.

Let's return to the twelfth story for a while. In the first section of this story we posed the problem of minimization or maximization of a function of a number of variables subject to equality and inequality constraints. In the absence of inequality constraints the problem would take the form

$$\text{(p)} \qquad F_0(x) \to \min(\max), \qquad F_i(x) = 0, \, i = 1, \ldots, m,$$

where $x = (x_1, \ldots, x_n)$ and the $F_i(x)$ are functions of n variables (we have deliberately replaced f_i by F_i).

Now let's study exclusively problems in which the F_i are not functions of many variables, but functionals like the $F(y)$ introduced earlier. Such functionals are called *functionals of the classical calculus of variations*. The preceding sections contain a number of relevant examples, such as the length functional $f(x, y, z) = \sqrt{1 + z^2}$, the area functional $f(x, y, z) = y$, and the Bernoulli functional $f(x, y, z) = \sqrt{1 + z^2}/\sqrt{2gy}$.

Let $f_0(x, y, z), f_1(x, y, z), \ldots, f_m(x, y, z)$ be a selection of functions. Consider their corresponding functionals $F_0(y), F_1(y), \ldots, F_m(y)$ from the calculus of variations and the following variational problem

$$\text{(p}_1) \qquad F_0(y) \to \min(\max), \qquad F_i(y) = \alpha_i, \, i = 1, \ldots, m.$$

This problem is called *the isoperimetric problem* of the classical calculus of variations. The functions $y(x)$ in $C^1([a, b])$ satisfying the conditions $F_i(y) = \alpha_i$, $i = 1, \ldots, m$, $y(a) = y_0$, and $y(b) = y_1$, are said to be *admissible in the problem* (p$_1$).

In the absence of constraints of the type of $F_i(y) = \alpha_i$, problem (p$_1$) takes the form

$$\text{(p}_2) \qquad F_0(y) \to \min(\max) \left(\Leftrightarrow \int_a^b f_0(x, y(x), y'(x))\,dx \to \min(\max) \right)$$

over all $y(x)$ such that $y(a) = y_0$, and $y(b) = y_1$. Problem (p$_2$) is called *the simplest problem* of the classical calculus of variations.

The brachistochrone problem belongs to this class of simplest problems. Dido's second problem is part of the class of isoperimetric problems (hence the term "isoperimetric" as applied to problem (p$_1$)).

Newton's problem belongs to neither of these classes, because the constraint $y'(x) \geq 0$ is absent from (p_1) as well as from (p_2).

Now arises the question of how to define the notion of a local minimum (maximum) in problem (p_1). To answer this question it is necessary to introduce some measure of "distance" between functions in $C^1([a, b])$. We take as the distance between a function $y(x)$ in $C^1([a, b])$ and the function that is identically zero the number

$$\|y\|_1 = \max_{x \in [a, b]} |y(x)| + \max_{x \in [a, b]} |y'(x)|$$

(called the *norm* of the function $y(x)$), and as the distance between functions $y_1(x)$ and $y_2(x)$ the number of $\|y_1 - y_2\|_1$. In the space $C([a, b])$, we define the norm of a function $y(x)$ as

$$\|y\|_0 = \max_{x \in [a, b]} |y(x)|.$$

Now we can define a local minimum in a way that is entirely analogous to definition 1 in the twelfth story.

DEFINITION. A function $\hat{y}(x)$ is said to *yield a local minimum (maximum) in problem* (p_1) if there is a $\epsilon > 0$ such that for all functions admissible in (p_1) and satisfying the inequality

$$\|y - \hat{y}\| < \varepsilon$$

we have the inequality $F_0(y) \geq F_0(\hat{y})(F_0(y) \leq F_0(\hat{y}))$.

Now we come to the key question of how to solve problem (p_1).

Recall the meaning of the Lagrange principle as applied to problem (p). It consisted of two assertions.

1. For unconstrained problems a necessary condition for an extremum at a point $\hat{x}$ is the equality

$$F_0'(\hat{x}) = 0,$$

(Fermat's theorem).

2. To solve problem (p) we must form the Lagrange function and treat it as if the variables were independent (that is, we must apply Fermat's theorem). We called the second assertion the Lagrange principle.

All this turns out to have a perfect analog in the case of problem (p_2). All we need do is modify the meaning of Fermat's theorem. Specifically, we have the following theorem.

THEOREM (Euler). *Let f_0 in the simplest problem (p_2) be a continuously differentiable function of three variables. If a function $\hat{y}(x)$ yields a local extremum (minimum or maximum) in the simplest problem (p_2) then the following equation holds*:

(1) $$\frac{d}{dx} f_{0y'}(x, \hat{y}(x), \hat{y}'(x)) - f_{0y}(x, \hat{y}(x), \hat{y}'(x)) = 0.$$

This equation is called Euler's equation for the problem (p_2). Its admissible solutions are called the *stationary* points or *extremals* of the problem.

Equation (1) is a decoded version of an equation of type $F_0'(\hat{x}) = 0$ as applied to the simplest problem. We will try to explain this in the next section. However, in order to solve concrete problems we need not know the origin of equation (1). Thus the algorithm (that goes back to Euler) for the solving of simplest problems consists of the following:

Find all solutions of equation (1) *(they depend on two variables) that pass through the given points and select the ones for which the functional* F_0 *takes on its least (largest) value.*

It is easy to show by direct differentiation that if f_0 does not depend on x, then equation (1) admits of the following relation ("integral"):

$$(1') \qquad f_0(\hat{y}(x), \hat{y}'(x)) - y'(x) f_{y'}(\hat{y}(x)), \qquad \hat{y}'(x)) \equiv \text{constant}.$$

In other words, every solution of equation (1) satisfies $(1')$.

The Lagrange method can be applied to the general problem (p_1) without any modification. We must form the Lagrange function

$$\mathcal{L} = \lambda_0 F_0(y) + \lambda_1 F_1(y) + \cdots + \lambda_m F_m(y),$$

which can also be written as

$$\mathcal{L} = \int_a^b f(x, y(x), y'(x))\, dx,$$

with

$$f(x, y, z) = \lambda_0 f_0(x, y, z) + \lambda_1 f_1(x, y, z) + \cdots + \lambda_m f_m(x, y, z),$$

and proceed as if our task was to find an extremum of the function $\mathcal{L}$ where the functions y are independent. In other words, we must write down the Euler equation

$$\frac{d}{dx} f_{y'} - f_y = 0 \Leftrightarrow \frac{d}{dx}(\lambda_0 f_{0y'} + \cdots + \lambda_m f_{my'}) - (\lambda_0 f_{0y} + \cdots + \lambda_m f_{my}) = 0.$$

All this is based on the following theorem.

THEOREM (the Lagrange multiplier rule for isoperimetric problems). *Let* $f_0, \ldots, f_m$ *be continuously differentiable functions. If a function* $\hat{y}(x)$ *yields a local extremum (minimum or maximum) in problem* (p_1), *then there are numbers* $\lambda_0, \ldots, \lambda_m$ *not all zero such that Euler's equation holds:*

$$\frac{d}{dx}(\lambda_0 f_{0y'}(x, \hat{y}(x), \hat{y}'(x)) + \cdots + \lambda_m f_{my'}(x, \hat{y}(x), \hat{y}'(x))$$

$$(2) \qquad - (\lambda_0 f_{0y}(x, \hat{y}(x), \hat{y}'(x)) + \cdots + \lambda_m f_{my}(x, \hat{y}(x), \hat{y}'(x))) = 0.$$

The admissible solutions of equation (2) are called *stationary solutions*.

The Lagrange multiplier rule justifies the following four-stage prescription for obtaining a solution of problem (p_1).

$1°$ Formalization of the problem. $2°$ Application of the Lagrange principle, that is, setting down equation (2) together with the equations $F_i(y) = \alpha_i$ and the boundary conditions $y(a) = y_0$, $y(b) = y_1$. $3°$ Finding all stationary solutions. $4°$ Selection of the stationary solutions that are solutions of the problem.

We'll use this algorithm to solve our two problems, namely the brachistochrone problem and Dido's problem.

Solution of the brachistochrone problem.

$1°$ *Formalization.* We formalized the brachistochrone problem as one of the simplest type with

$$f_0(x, y, z) = \sqrt{1 + z^2} / \sqrt{2gy},$$

(f_0 not dependent on x).

$2°$ *Necessary condition.* Euler's equation

$$\frac{d}{dx} f_{0y'} - f_{0y} = 0,$$

admits the integral

$$f_0 - y' f_{0y'} = \text{constant} \Leftrightarrow \frac{\sqrt{1 + y'^2}}{\sqrt{2gy}} - \frac{y'^2}{\sqrt{1 + y'^2}\sqrt{2gy}} = \text{constant}$$

$$(*) \qquad\qquad \Rightarrow \sqrt{1 + y'^2}\sqrt{y} = C,$$

where C is some constant. We recall that this very relation was obtained by Johann Bernoulli.

$3°$ *Finding the stationary points.* This is tantamount to finding the solutions of equation $(*)$. But we have already integrated this equation and found that its solutions are a family of cycloids.

$4°$ *Discussion.* It was shown in the seventh story that there is just one admissible cycloid in the family of cycloids that are the solutions of $(*)$. This cycloid is the solution of the problem (of course, this assertion requires justification).

Solution of Dido's second problem.

$1°$ *Formalization.* We've already formalized Dido's problem as an isoperimetric problem with $f_0(x, y, z) = y$, $f_1(x, y, z) = \sqrt{1 + z^2}$.

$2°$ *Necessary condition.* The Lagrange principle.   We form the sum

$f = \lambda_0 f_0 + \lambda_1 f_1$ and write down Euler's equation

(**)
$$\frac{d}{dx} \lambda_1 \frac{y'}{\sqrt{1+y'^2}} - \lambda_0 = 0.$$

3° *Finding the stationary points.* We solve this equation under the assumption that if $\lambda_0 = 0$, then $\hat{y} \equiv 0$ (check this bearing in mind the boundary conditions). This is possible only if $l = a$. Thus if $l > a$, then we can assume that $\lambda_0 = 1$. Then (**) yields

$$\frac{d}{dx} \lambda_1 \frac{y'}{\sqrt{1+y'^2}} = 1 \Rightarrow \frac{\lambda_1 y'}{\sqrt{1+y'^2}} = x + c \Rightarrow \frac{y'}{\sqrt{1+y'^2}} = Cx + D$$

$$\Rightarrow \frac{y'^2}{1+y'^2} = (Cx+D)^2 \Rightarrow \frac{dy}{dx} = \frac{\pm(Cx+D)}{\sqrt{1-(Cx+D)^2}}$$

$$\Rightarrow dy = \pm \frac{(Cx+D)\,dx}{\sqrt{1-(Cx+D)^2}} \Rightarrow d\left(y \pm \frac{1}{C}\sqrt{1-(Cx+D)^2}\right) = 0$$

$$\Rightarrow (x+a)^2 + (y+b)^2 = r^2.$$

This is the family of all circles.

4° *Discussion.* Now it is easy to find the required solution. If $a < l \leq (\pi a)/2$, then our family of circles contains just one circle with perimeter l passing through the points $(0,0)$ and $(a,0)$. If $l > (\pi a)/2$, then the solution will be the semicircle with center $(a/2, (l-(\pi a)/2)$ and radius $a/2$ "supplemented" by the segments $x = 0$, $0 \leq y \leq l - (\pi a)/2$, $x = a$, $0 \leq y \leq l - (\pi a)/2$. (See Figure 14.1 on page 148.)

Note the difference between our solutions in the thirteenth story and here. In the earlier story the problems were solved "to the very end." Here there is an element of indeterminacy connected with the existence of a solution. In problems of the calculus of variations, existence of solutions is more difficult to establish than in the finite-dimensional case. Also, in the calculus of variations it is often the case that solutions just don't exist. For example, in the just-investigated problem of Dido, there is no solution in the usual sense for $l > (\pi a)/2$; indeed, the raised semicircle "supplemented by segments" is not a function that joins the points $(0,0)$ and $(a,0)$. In such cases mathematicians speak of "generalized" solutions.

In the case of the brachistochrone, matters are not as simple as they may at first appear. The difficulty is that a cycloid is not continuously differentiable. In other words, there is no solution of the brachistochrone problem in the totality of functions (in the space $C^1([a, b])$) in which the problem was considered.

The eminent twentieth-century mathematician David Hilbert (whose words serve as an epigraph for this story) advanced the view that every reasonable variational problem must have a solution "if, whenever necessary, the notion of a solution is given an extended meaning."

Hilbert's idea turns out to be correct for most problems, including the brachistochrone problem, Dido's problem, and the two problems formulated at the end of the seventh story. We will now turn to the solution of these two problems.

L'Hospital's problem

$1°$ *Formalization.* The time of propagation of light from the point $(0, y_0)$ to the point (a, y) in a medium in which the velocity of propagation depends only on the altitude y and is equal to $v(y)$ is given by the integral

$$T(y) = \int_0^a \frac{\sqrt{1 + (y'(x))^2}}{v(y)} \, dx.$$

To see that this is so, it suffices to take another look at formula (2) in the seventh story. In sum, we end up with the following simplest problem of the classical calculus of variations:

$$\int_0^a \frac{\sqrt{1 + (y'(x))^2}}{y} \, dx \to \min,$$

$$y(0) = y_0, \qquad y(a) = y_1 \qquad \left(f_0(x, y, z) = \frac{\sqrt{1 + z^2}}{y} \right).$$

$2°$ *Necessary condition.* Euler's equation admits the following integral:

$$f_0 - y' f_{0y'} = \text{const} \Leftrightarrow y\sqrt{1 + y'^2} = D^2.$$

$3°$ *Finding the stationary points.*

$$y\sqrt{1 + y'^2} = D^2 \Rightarrow \frac{y \, dy}{\sqrt{D^2 - y^2}} = dx \Rightarrow x - C_1 = \int \frac{y \, dy}{\sqrt{D^2 - y^2}}$$

$$\Rightarrow (x - C_1)^2 + y^2 = C^2.$$

We've integrated Euler's equation. The result is a family of semicircles with centers on the x-axis.

$4°$ *Discussion.* It is easy to see that for any two points $(0, y_0)$ and (a, y_1) there is exactly one circle from our family of circles that passes through those points. This circle is the solution of our problem. A proof of this fact is beyond the scope of this book. *It is interesting that these very semicircles are straight lines in the Poincaré model of the hyperbolic plane.*

Problem of the minimal surface of revolution

$1°$ *Formalization.*

$$\int_{x_0}^{x_1} y\sqrt{1 + (y'(x))^2}\, dx \to \min, \qquad y(x_0) = y_0, \qquad y(x_1) = y_1.$$

Recall that, except for the missing factor 2π, this "area-of-a-surface-of-revolution" functional was introduced at the end of the seventh story. Thus, here

$$f_0(x, y, z) = y\sqrt{1 + z^2}.$$

$2°$ *Necessary condition.* Euler's equation admits the following integral:

$$f_0 - y'f_0 = \text{const} \Rightarrow \frac{1 + y'^2}{y^2} = D^2 \Rightarrow \frac{dy}{\sqrt{D^2 y^2 - 1}} = dx.$$

$3°$ *Finding the stationary points* (*extremals*). We solve the equation in Section 2 with the help of a substitution:

$$Dy = (e^t + e^{-t})/2 \Rightarrow D\, dy = (e^t - e^{-t})\, dt/2,$$

$$\sqrt{D^2 y^2 - 1} = \sqrt{(e^t + e^{-t})^2/4 - 1} = (e^t - e^{-t})/2$$

$$\Rightarrow \frac{dy}{\sqrt{D^2 y^2 - 1}} = dt \Rightarrow dt = dx \Rightarrow t = x + D_1,$$

$$y = (e^{Dx + D_1} + e^{-(Dx + D_1)})/D.$$

The curve

$$y = (e^{Dx} + e^{-Dx})/2$$

is called *a catenary*.

$4°$ *Discussion.* We have shown that if a solution of the problem of a minimal surface exists, then the curve of revolution is a catenary.

At this point we will again discuss Hilbert's idea of a generalized solution. In the brachistochrone problem and in l'Hospital's problem, we obtained solutions. These are "genuine," rather than generalized, solutions. True, in both cases, the extremals are not continuously differentiable functions when the ordinate of one of the endpoints is zero. (Note that in l'Hospital's problem, light from such a point would take an "infinite time" to propagate, so that there are physical reasons for ignoring points with zero ordinate.)

In Dido's problem we obtained a solution for $l \le \pi/2$ and a generalized solution for $l > \pi/2$. In the minimal-surface problem the situation is more complex in the sense that sometimes a classical solution exists, and sometimes it doesn't (see Figure 14.2, where $x_0 = -a$, $x_1 = a$, $y_1 = y_0$). If there is no

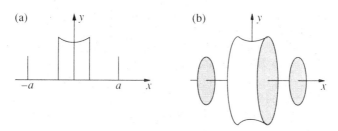

FIGURE 14.2

classical solution, then the minimum is yielded by the generalized solution consisting of the segments $x = -a$, $0 \le y \le y_0$; $x = a$, $0 \le y \le y_1$, joined by the segment $y = 0$, $-a \le x \le a$. (See Figure 14.2(a).) Then the "surface of revolution" consists of two disks connected by the "bridge" $y = 0$, $-a \le x \le a$. (See Figure 14.2(b).) When a solution exists, we must choose the minimum from the classical and generalized solutions.

4. From the history of the calculus of variations. In the previous section we explained the substance behind the term "calculus of variations." Now the time has come to discuss the historical evolution of this discipline and the origin of the term itself.

We recall that it all began with the brachistochrone—the problem posed in 1696 by Johann Bernoulli. This problem attracted universal attention, and soon a few similar problems were solved. (We dealt with some of them, namely l'Hospital's problem and the minimal-surface problem.) While each problem was solved individually, it was sensed that a uniform approach is possible.

Then Johann Bernoulli set his student Leonhardt Euler the task of trying to find a general method for solving all such problems. Euler succeeded. In 1744 Euler published the memoir, *A method for discovering curved lines having a maximum or minimum property or the solution of the isoperimetric problem taken in its widest sense.* Euler's method involved finding the equation that must be satisfied by a "curved line having a maximum or minimum property." We set this equation down before; it came to be known as *Euler's equation.*

Note the term "isoperimetric" in the title of Euler's work. We began this book with the isoperimetric problem. How does it relate to Euler? It is true that by means of his method Euler was also able to solve the isoperimetric problem. But in numerous other cases that Euler was able to settle, the constraints had nothing to do with the length of the curve, with its perimeter. Nevertheless the term "isoperimetric" reflected the continuity of our discipline and the name stuck.

In 1759 a highly significant event occurred. The very young Lagrange wrote his work bearing on this topic. He approached it from a different direction, and with such success that henceforth his method (sometimes called the variational method) was universally adopted. Euler was delighted with Lagrange's paper and refrained from publishing his own elaborations on this topic so as to enable the young scholar to carry his designs to completion. Euler called the whole new chapter of mathematics *the calculus of variations*.

What is the essence of Lagrange's method? What is a variation?

Let's return to Section 2 of the twelfth story where we derived the finite-dimensional version of Fermat's theorem. We then reasoned as follows.

Let $f_0(x)$ be a function of n variables $x = (x_1, \ldots, x_n)$. We assume that the function is differentiable and that it attains a local extremum at a point $\hat{x}$. Then it is clear that the function of one variable

$$g(\lambda) = g(\lambda, j) = f_0(\hat{x}_1, \ldots, \hat{x}_{j-1}, \hat{x}_j + \lambda, \hat{x}_{j+1}, \ldots, \hat{x}_n)$$

must have a minimum at zero. In view of the (one-dimensional) theorem of Fermat, the following equality must hold

$$g'(0) = 0 \Leftrightarrow \frac{\partial f(\hat{x})}{\partial x_j} = 0.$$

Lagrange applied this very method to the simplest problem of the classical calculus of variations:

(1) $F(y) = \displaystyle\int_{x_0}^{x_1} f(x, y(x), y'(x))\, dx \to \min(\max),\ y(x_0) = y_0,\ y(x_1) = y_1.$

We will follow Lagrange's train of thought. Suppose that the function $f(x, y, z)$ in (1) is continuously differentiable and the functional $F(y)$ attains a local minimum for the continuously differentiable curve $\hat{y}(x)$. Now we take a "variation" of $\hat{y}(x)$. Specifically, we take any continuously differentiable curve $y(x)$ that vanishes at the endpoints: $y(x_0) = y(x_1) = 0$. Then "variation" of $\hat{y}(x)$, that is, addition to $\hat{y}(x)$ of the function $y(x)$ multiplied by any number λ, does not take us outside the set of admissible curves—indeed, all of them pass through the points (x_0, y_0) and (x_1, y_1). This means that the function of one variable

$$g(\lambda) = g(\lambda, y) = F(\hat{y} + \lambda y) = \int_{x_0}^{x_1} f(x, \hat{y}(x) + \lambda y(x), \hat{y}'(x) + \lambda y'(x))\, dx$$

must have a local minimum at zero. But then we can again apply the one-dimensional Fermat theorem. This theorem implies that $g'(0) = 0$. Now we compute $g'(0)$.

One proves in analysis that our assumptions about the continuous differentiability of $f(x, y, z)$ and $y(x)$ justify differentiation under the integral sign. After some simple computations we find that

(2) $g'(0) = \displaystyle\int_{x_0}^{x_1} (a(x)y(x) + b(x)y'(x))\, dx,$

where $a(x)$ and $b(x)$ denote $\frac{\partial f}{\partial y}(x, \hat{y}(x), \hat{y}'(x))$ and $\frac{\partial f}{\partial y'}(x, \hat{y}(x), \hat{y}'(x))$ respectively. Thus, following Lagrange, we conclude that if $\hat{y}(x)$ yields a local maximum or minimum in the simplest problem (1), then, for any continuously differentiable function $y(x)$ with $y(x_0) = y(x_1) = 0$, we have the equality

(3)
$$\int_{x_0}^{x_1} (a(x)y(x) + b(x)y'(x))\, dx = 0.$$

We will continue our reasoning. We find a function $c(x)$ such that $c'(x) = a(x)$ and $\int_{x_0}^{x_1} c(x)\, dx = \int_{x_0}^{x_1} b(x)\, dx = B$. To this end we choose a constant D such that the integral of the function $c(x) = \int_{x_0}^{x} a(\xi)\, d\xi + D$ over the interval $[x_0, x_1]$ is B. Integrating the first summand in (3) by parts, we find that

$$0 = \int_{x_0}^{x_1} (a(x)y(x) + b(x)y'(x))\, dx = \int_{x_0}^{x_1} (c'(x)y(x) + b(x)y'(x))\, dx$$

$$= \int_{x_0}^{x_1} (b(x) - c(x))y'(x)\, dx.$$

Now we take the last step. We put $\bar{y}(x) = \int_{x_0}^{x} (b(\xi) - c(\xi))\, d\xi$. Then it is clear that $\bar{y}(x_0) = 0$. Also, $\bar{y}(x_1) = \int_{x_0}^{x_1} (b(x) - c(x))\, dx = 0$ by the construction of the function $c(x)$. Hence $\bar{y}'(x) = b(x) - c(x)$; this follows from the Newton-Leibniz formula. But then (3) must hold for $\bar{y}(x)$, that is, $\int_{x_0}^{x_1} (b(x) - c(x))^2\, dx = 0$. Since the integral of a continuous positive function cannot be zero, we conclude that $b(x) \equiv c(x)$. This means that

$$b'(x) = a(x),$$

that is,

(4)
$$\frac{d}{dx}\frac{\partial f}{\partial y'}(x, \hat{y}(x), \hat{y}'(x)) - \frac{\partial f}{\partial y}(x, \hat{y}(x), \hat{y}'(x)) \equiv 0.$$

This is the derivation of Euler's equation in the manner of Lagrange.

Recall that in the previous section we said that Euler's equation is the decoded version of Fermat's theorem for the simplest problem. By now the meaning of this remark is clear. Euler's equation is a consequence of the fact that *the derivative of the functional $F(y)$ at the point $\hat{y}(x)$ in every direction $y(x)(y(x_0) = y(x_1) = 0)$ is zero*. We note that the expression for $g'(0)$ has come to be known as *the variation of the functional F*.

Lagrange did not stop with the mathematical problems of the calculus of variations. Strictly speaking, he concerned himself with these problems so that he could apply them to problems in the natural sciences. His life's main work—*Analytical mechanics*—is a book about motions of physical bodies. At the basis of Lagrange's approach to mechanics lies an extremal principle known as *the principle of least action.* We will illustrate it using a very simple example.

Suppose a small ball of mass m is attached to a spring of negligible weight that obeys Hooke's law of the proportionality of the tension in the spring and its deflection from the rest position O. The spring is aligned with the y-axis. The displacement of the ball is given by a function $y(t)$, where $y(t)$ is the ball's coordinate at time t. It is well known that $y(t)$ satisfies Newton's law

$$(5) \qquad\qquad y''(t) = -k\, y(t)$$

which asserts that "*the product of mass by acceleration equals the acting force*" (the acceleration is given by the second derivative and the force, by Hooke's law, equals $-k\, y(t)$, where k is a proportionality constant). In mechanics $T = m(y'(t))^2/2$ is called the *kinetic energy*, $U = ky^2(t)/2$ the *potential energy*, and the integral of the difference of the kinetic and potential energies is called *the action.*

Let's consider the problem of minimizing the action for fixed boundary conditions:

$$(6)$$
$$\int_{t_0}^{t_1} (T - U)\, dt = \int_{t_0}^{t_1} \left(\frac{my'^2}{2} - \frac{ky^2}{2} \right) dt \to \min, \quad y(t_0) = y_0, \quad y(t_1) = y_1.$$

Euler's equation for problem (6) yields equation (5):

$$f = \frac{my'^2}{2} - \frac{ky^2}{2} \Rightarrow \frac{\partial f}{\partial y'} = my',$$

$$\frac{\partial f}{\partial y} = -ky \Rightarrow \frac{d}{dt} f_{y'} - f_y = 0 \Leftrightarrow my'' + ky = 0.$$

Thus *Newton's second law is none other than Euler's equation for the action.* Put differently, *the actual motions are determined by the stationary points of the action.* For small time intervals the actual trajectory does indeed minimize the action, so that the *principle of least action* is true for such intervals. In general, it is more appropriate to speak of the *principle of stationary action.*

We have again run up against the fact that the laws of nature admit dual descriptions, one "physical" and the other "extremal." This was first mentioned in the third story and recalled in the seventh. There we discussed optics and minimized time; here we describe the motion of bodies and minimize the action. Following Hamilton, who investigated optical phenomena from this point of view, Jacobi suggested the consideration of analogs of wavefronts in mechanical problems and, in general, in all of the simplest problems of the classical calculus of variations.

Jacobi considered the endpoint function $S(x, y)$, whose value is that of the integral $\int_{x_0}^{x} f(\xi, y(\xi))\, d\xi$ on the extremal (that yields a minimum) joining a fixed point (x_0, y_0) to the point (x, y).

It is obvious that *any part of an extremal that yields a miminum is an extremal that yields a minimum*. This simple observation is the decoding, in the general situation, of Huygens' principle in optics (mentioned in the third story) and is also called *Huygens' principle*. Using Huygens' principle and Euler's equation it is easy to derive the equation satisfied by the function $S(x, y)$. This equation is called the *Hamilton-Jacobi* equation. It has been possible to integrate this equation in many cases of interest. This method affords another possibility of investigation of problems of the classical calculus of variations. Thus the duality of the description of optical phenomena has led to a dual description of the solutions of an arbitrary problem of the classical calculus of variations.

Now we will go back in time, back to the eighteenth century. In order to be able to extract consequences from the principle of least action, it was necessary to learn to solve problems of the calculus of variations under constraints more complicated then isoperimetric constraints. Specifically, one had to learn to solve problems subject to constraints given by differential equations. Let's look at one general formulation (that goes back to Lagrange) to which the majority of the most interesting applied problems can be reduced. Let $f_j = f_j(x, y_1, \ldots, y_n, z_1, \ldots, z_n)$, $j = 0, 1, \ldots, k$, $k < n$, be functions of $2n + 1$ variables. We consider the problem

$$F_0(y) = \int_{x_0}^{x_1} f_0(x, y_1, \ldots, y_n, y_1', \ldots, y_n') \, dx \to \min(\max),$$

$$f_1(x, y_1, \ldots, y_n, y_1', \ldots, y_n') = 0,$$

$$\cdots \cdots \cdots \cdots \cdots \cdots \cdots \cdots \cdots \cdots$$

$$f_k(x, y_1, \ldots, y_n, y_1', \cdots, y_n') = 0$$

with boundary conditions $y_i(x_0) = y_{i0}$, $y_i(x_1) = y_{i1}$, $i = 1, \ldots, n$. This is the so-called *Lagrange problem*. How does one write down for it the necessary extremum conditions?

Lagrange was convinced that this problem is also governed by the principle of "lifting the constraints" that we discussed in the twelfth story (the Lagrange principle). In accordance with Lagrange's general conception, we must form a Lagrange function and set down the necessary condition for an extremum problem for the Lagrange function in the absence of constraints. But what does the Lagrange function for the Lagrange problem look like? Here, too, Lagrange displayed supreme decisiveness. In the finite-dimensional case, it is the sum of a functional multiplied by a number λ_0 and of products of the constraint functions by the Lagrange multipliers. But in the case of the Lagrange problem "there are as many constraints as points in the interval $[x_0, x_1]$." This being so, Lagrange proposed to multiply the ith equation $f_i(x, y_1(x), \ldots, y_n(x), y_1'(x), \ldots, y_n'(x)) = 0$ by a function $l_i(x)$ of x, integrate over the interval $[x_0, x_1]$, and, finally, sum over i. In other words,

Lagrange replaced multiplication by numbers followed by summation with multiplication by functions followed by integration. All in all, the Lagrange function took the following form:

$$\mathscr{L} = \int_{x_0}^{x_1} f(x, y_1(x), \ldots, y_n(x), y_1'(x), \ldots, y_n'(x))\, dx,$$

where

$$f = \lambda_0 f_0 + \sum_{i=1}^{k} l_i(x) f_i.$$

Then Lagrange formulated the following result: If certain functions $\hat{y}(x) = (\hat{y}_1(x), \ldots, \hat{y}_n(x))$ yield a local minimum of the Lagrange problem, then there are a number λ_0 and functions $l_i(x)$ such that Euler's equation holds for the function f. (Of course, Lagrange multiplied the functional by 1 rather than by λ_0).

Lagrange did not prove his result. Of course, a result formulated without any restrictions cannot be true. Thus the very method underlying his supreme work—his *Analytical mechanics*—lacked a rigorous justification. This state of affairs continued for over a century. A completely rigorous proof of Lagrange's theorem was given only at the end of the nineteenth century, and its essence was understood only in our own century. The meaning of "was understood only in our own century" deserves a special discussion that is the subject of the next section.

5. Conclusion. In this section we'll say more about infinite-dimensional and convex analysis, the theory of optimal control and Pontryagin's maximum principle, the rapid-response problem, and Newton's problem.

This book is meant for high school students. In Part One I refrained from introducing any element of mathematical analysis. In Part Two we discussed things not covered in school, such as functions of more than one and even of infinitely many variables. Nevertheless, I have stayed pretty close to the high school curriculum. And even in this concluding section of my concluding mathematical story—the fifteenth story is given over to general questions and free-wheeling talk—I am reluctant to abandon my approach of talking "to the first high school student in the street" (recall Hilbert's words). But I'd like to be less constrained and, in my mind, address only such a "first high school student in the street" who has decided to tie his or her future to mathematics. I intend to speak as if I could predict the future. I count on the student to eventually fill in the gaps in understanding that are due to the limitations of his or her present knowledge.

I will also point out and partially justify certain general theses pertaining to the subsequent fate of the ideas discussed earlier. They seem destined to evolve and also to return to their sources.

You will recall that the method of investigation of maximum and minimum problems was elaborated first by Fermat (for polynomials) and then, in general terms, by Newton and Leibniz. Immediately thereafter began the period of development of the classical calculus of variations, of investigation of extrema of certain functions of infinitely many variables. This period lasted for about two-and-a-half centuries.

At the end of the nineteenth century Volterra, and somewhat later Fréchet, Hadamard, and many others, began to develop the foundations of infinite-dimensional analysis. In this connection it was stressed that one of the aims of the newly created calculus was the solution of maximum and minimum problems (recall the words of Volterra that form one of the epigraphs for this story). In the first half of this century, mathematical analysis in infinite-dimensional spaces (now called *functional analysis*, this chapter of analysis has unified various conceptions of classical analysis, higher algebra, and geometry) experienced a period of explosive development and growth. But the mathematicians who continued to develop the calculus of variations at that time did not apply the general theorems of functional analysis to the calculus of variations and, furthermore, were unaware that what was being elaborated was an apparatus for this theory. In his textbook on the calculus of variations published in the 1940s, in which he summarized the whole development of this discipline at a time when all the necessary results of functional analysis were already common knowledge, George Bliss, one of this century's foremost experts on the calculus of variations, spoke in highly skeptical terms of the potential utility of this general approach for the subject's "concrete" problems.

At this point it is perhaps time to formulate our first thesis.

Infinite-dimensional analysis (more precisely, the differential and integral calculus in infinite-dimensional spaces), *a division of mathematics based on exactly the same ideas as finite-dimensional analysis and just as simple and natural as the latter, provides as natural an apparatus for the classical calculus of variations as does finite-dimensional analysis for the theory of finite-dimensional extremal problems.* Also, the fundamental theorems of the differential and integral calculus in infinite-dimensional spaces are just as simple and natural as their finite-dimensional analogs.

The basic concepts of the differential and integral calculus of functions of one variable are *the derivative and the differential*. We defined them first in the eleventh story.

We say that a function $F(x)$ defined on the real line $\mathbb{R}$ is *differentiable at a point* x_0 if there is a linear function $y = kx$ such that $F(x_0+x) - F(x_0) = kx + r(x)$, where $\lim_{|x| \to 0} |r(x)|/|x| = 0$.

The linear function $y = kx$ is called *the differential of F at the point x_0*.

Infinite-dimensional analysis is concerned with functions defined on *spaces with a norm*, known as *normed spaces*. Examples of normed spaces include the spaces $C([a, b])$ and $C^1([a, b])$ that we encountered earlier. More generally, a normed space is any set Y of elements y that can be handled like plane vectors, that is, that can be added and multiplied by numbers, and, in addition, are each assigned a number $\|y\|$ such that

(1) $\|y\| \geq 0$ (nonnegativity) and $\|y\| = 0$ only if $y = 0$,
(2) $\|ay\| = |a|\|y\|$ for all $y \in Y$ and $a \in \mathbb{R}$ (homogeneity),
(3) $\|y_1 + y_2\| \leq \|y_1\| + \|y_2\|$ for all $y_1, y_2 \in Y$ (the triangle inequality).

Other examples of normed spaces are the line $\mathbb{R}$ with $\|y\| = |y|$ and the n-dimensional space of vectors $y = (y_1, \ldots, y_n)$ with various norms, exemplified by

$$\|y\| = |y_1| + \cdots + |y_n|, \quad \text{and} \quad \|y\| = \sqrt{y_1^2 + \cdots + y_n^2},$$

and the spaces C and C^1 above.

I have already mentioned that a function $K(y)$ is said to be *linear* if $K(y_1 + y_2) = K(y_1) + K(y_2)$ for all $y_1, y_2 \in Y$ and $K(ay) = \alpha K(y)$ for all $y \in Y$ and $a \in \mathbb{R}$.

We can now give a general definition of a derivative.

We say that a function $F(y)$ defined on a normed space Y is differentiable at a point y_0 if there is a linear function $K(y)$ such that

$$F(y_0 + y) - F(y_0) = K(y) + r(y),$$

where $\lim_{\|y\| \to 0} |r(y)|/\|y\| = 0$.

The linear function $K(y)$ is called *the differential of F at the point y_0*.

Don't you agree that this is much the same thing? That's not all. The definition of a differential in infinite-dimensional analysis was given in the beginning of this century by the French mathematician M. Fréchet. Let's see what this definition leads to in the finite-dimensional case.

We will say that a function $F(y) = F(y_1, \ldots, y_n)$ of n variables is *differentiable* at a point $y_0 = (y_{01}, \ldots, y_{0n})$ if there is a linear function (recall that any such function has the form $K(y) = k_1 y_1 + \cdots + k_n y_n$) such that

$$F(y_{01} + y_1, \ldots, y_{0n} + y_n) - F(y_{01}, \ldots, y_{0n}) = K(y) + r(y),$$

where $\lim_{\|y\| \to 0} |r(y)|/\|y\| \to 0$. As norm we can take any norm in n-dimensional space.

Nowadays this definition appears in every textbook on analysis. But Fréchet believed that he was the first to think of it! That's how he puts it: "the differential in my sense of the term." Ordinary finite-dimensional analysis existed for two and a half centuries before him and yet this top mathematician of the beginning of this century thought that he was the first to proffer the correct definition of the fundamental concept of analysis—that of a differential! True, it then turned out that "his" definition had already been given by Weierstrass in unpublished works (written in the 1860s) and

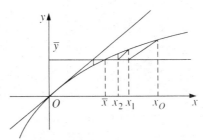

FIGURE 14.3

is found in early twentieth-century English and German textbooks, but the definition was actually absent from the scientific literature. The fact remains that it was easier to conceive the infinite-dimensional definition than the finite-dimensional one!

We go on. What are the fundamental theorems of the differential calculus? We will begin with two, namely the chain rule and the inverse function theorem.

The chain rule (for a function of one variable) states that

$$(F(G(x)))' = F'(G(x))G'(x).$$

When appropriately interpreted, this formula holds in infinite-dimensional analysis as well. It is proved with equal ease in both cases.

The inverse function theorem (for a function of one variable) states that if $y = f(x)$ is a continuously differentiable function such that $f(0) = 0$ and $f'(0) \neq 0$, then f is invertible near zero, that is, for any small enough $\overline{y}$ there is a unique $\overline{x}$ such that $\overline{y} = f(\overline{x})$.

How is this theorem proved? Of the many different proofs I prefer the one that goes back to Newton. It consists in constructing a sequence $\{x_n\}_{n \geq 0}$ that converges stepwise to $\overline{x}$. This sequence is constructed according to the following rule:

(1) $$x_{k+1} = x_k + (f'(0))^{-1}(\overline{y} - f(x_k)), \qquad k \geq 0,$$

and is represented geometrically in Figure 14.3. The zero-th approximation x_0 is taken arbitrarily, but sufficiently close to zero.

The same result holds in the infinite-dimensional case. I'll risk formulating it. Here it is not enough to have a norm. Another requirement is the so-called completeness property, "the absence of gaps" (the rational numbers don't form a complete space because they have gaps like $\sqrt{2}$, which is not rational). A complete normed space is called a Banach space.

Now let X and Y be two Banach spaces and let F be a function that maps X into Y. Its derivative is a continuous linear mapping of X into Y. The infinite-dimensional version of our "one-dimensional statement" $f'(0) \neq 0$ is that the derivative $F'(0)$ is invertible—that is, $F'(0)^{-1}$ is also a continuous linear mapping of Y into X). And now the inverse function theorem takes an entirely analogous form: if X and Y are two Banach spaces and $y = F(x)$ is continuously differentiable function such that $F(0) = 0$ and $F'(0)$ is invertible, then for every $\bar{y}$ close to zero there is a unique $\bar{x}$ such that $F(\bar{x}) = \bar{y}$.

The proof is basically unchanged: The proof of the convergence of the x_k defined by formula (1) to $\bar{x}$ is entirely analogous to the one on the line $\mathbb{R}$.

In summary, the fundamental concept of a differential, the formulations of the basic theorems and their meaning and proofs—all these are basically the same in the one-dimensional and the infinite-dimensional cases. Also, *the Lagrange principle holds in infinite-dimensional analysis. Its foundations are the same as in the finite-dimensional case and the relevant proofs are just as simple.* The details follow.

Let X and Y be Banach spaces. Let f_i be functionals on X, $i = 0, 1, \ldots, m$, and let F be a mapping from X into Y. We consider the problem

(p) $f_0(x) \to \min(\max)$, $F(x) = 0$, $f_i(x) = 0$, $i = 1, \ldots, m$.

In the absence of the mapping F and the functions f_i we obtain the unconstrained problem

(p$'$) $f_0(x) \to \min(\max)$.

As applied to problem (p), the Lagrange principle is formulated using the very same words that are used in the finite-dimensional case and in the case of the isoperimetric problems of the classical calculus of variations, namely:

1. For problem (p$'$) the necessary extremum condition is the relation

$$f_0'(\hat{x}) = 0,$$

(Fermat's theorem).

2. To solve problem (p) we must *form the Lagrange function of the problem and treat it as if the variables were independent.*

The standard form of the Lagrange function for problem (p) is

$$\mathcal{L} = \mathcal{L}(x, \lambda_0, \ldots, \lambda_m, \Lambda) = \lambda_0 f_0(x) + \cdots + \lambda_m f_m(x) + \Lambda(F(x)),$$

where $\lambda_0, \ldots, \lambda_m$ are numbers and $\Lambda(y)$ is a linear function on Y.

Thus, if $\hat{x}$ is a local extremum, then there are numbers $\lambda_0, \ldots, \lambda_m$ and a linear function $\Lambda(y)$ such that Fermat's theorem holds for the Lagrange function:

$$\frac{\partial \mathcal{L}}{\partial x}(\hat{x}, \lambda_0, \ldots, \lambda_m, \Lambda) = 0.$$

But this is how one formulates the Lagrange multiplier rule in the finite-dimensional case, as witness the corresponding theorem in the twelfth story. The just-formulated theorem on the Lagrange multiplier rule was proved in 1934 by the Soviet mathematician L. A. Lyusternik.

It must be added that in the infinite-dimenisonal case, in addition to the smoothness of the functions and mappings involved in the statement of the problem (this was the only requirement in the finite-dimensional case), there are two more requirements.

The first requirement is that the spaces X and Y must be complete, that is, they must be Banach spaces. The second is that the mapping F have certain special properties usually referred to as regularity properties (a sufficient condition for regularity, and thus for the validity of the Lagrange multiplier rule, is that the derivative $F'(x)$ maps X onto Y). The infinite-dimensional result is proved as simply as the finite-dimensional one. A reader who wishes to verify this should consult the book [2R] (in Russian) where both the finite-dimensional and the infinite-dimensional cases are proved. The proofs follow strictly parallel paths except that occasionally certain well-known results of advanced algebra and classical analysis are replaced by their generalizations in functional analysis (these generalizations are part of the irreducible minimum of university training in mathematics).

What are *convexity and convex analysis*? We learn about convexity in high school geometry. Recall that *a figure* in the plane or in space *is said to be convex* if together with any two of its points it contains the segment joining them; *a function is said to be convex* if its graph lies not higher than the chord joining any two points on that graph. Figure 11.1 shows various convex and nonconvex figures. A triangle is always convex. There are nonconvex quadrilaterals. Linear ($y = ax$, $y = a_1x_1 + \cdots + a_nx_n$) and affine ($y = ax + b$, $y = a_1x_1 + \cdots + a_nx_n + b$) functions are convex; of the quadratic trinomials $y = ax^2 + bx + c$, $a \neq 0$, only those with $a > 0$ are convex. Convex functions can be defined analytically. A function $y = f(x)$ is said to be convex if and only if for any two points x_1 and x_2 and any number α between 0 and 1, we have Jensen's inequality

$$f(\alpha x_1) + (1 - \alpha)x_2) \leq \alpha f(x_1) + (1 - \alpha)f(x_2).$$

This inequality can be extended to the case of n points:

$$f(\alpha_1 x_1 + \cdots + \alpha_n x_n) \leq \alpha_1 f(x_1) + \cdots + \alpha_n f(x_n)$$

(provided that $\alpha_1 \geq 0, \cdots, \alpha_n \geq 0, \sum_{i=1}^{n} \alpha_i = 1$).

The following is an important convexity criterion for functions of one variable: if a function is twice differentiable and $f''(x) \geq 0$ for all x then it is convex. Now let's look at the functions in the table given in the eleventh story. It is easy to see that the function $y = |x|^a$ is convex if and only if $a \geq 1$, the function $y = a^x (a > 0$ and $\neq 1)$ is always convex, and so are the functions $y = \log_a x$, $0 < a < 1$, and $y = -\ln x$. The functions $y = \sin x$ and $y = \cos x$ are not convex.

Convex figures form a rather narrow and specialized class in the totality of figures. A similar statement is true of convex functions. But convexity plays

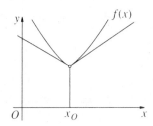

FIGURE 14.4

a very important part in mathematics and in its applications. The interesting ideas associated with convexity, and the wealth of applications, have led to the creation of a chapter of mathematics called "convex analysis." Its final formation took place relatively recently, about 20 years ago. In this chapter of mathematics one studies properties of convex sets, convex functions, and convex extremal figures. Of course, in this book we are primarily interested in convex extremal problems.

It was precisely the abundance of convex extremal problems that inevitably led to a deep study of convexity and, in effect, to the creation of convex analysis. The number of convex problems is particularly large in economics. We've already talked about one economic problem, namely the transportation problem. Such problems arise constantly in real life. Very frequently, the formalization of such problems has shown that the functions to be maximized or minimized, as well as the functions defining the constraints (of the types of equality and inequality), are linear. The methods for solving these problems form a special chapter of convex analysis known as linear programming. The first papers dealing with this field are due to the Soviet mathematician, academician L. V. Kantorovič, winner of the Lenin and Nobel prizes.

What is studied in convex analysis? One important area is the so-called "convex calculus" that has much in common with the differential calculus. We'll explain what the two calculi have in common.

Not all convex functions are differentiable. An example of a convex function that is not differentiable at zero is the function $y = |x|$. We have already seen that this function has no tangent (at zero). But every convex function $y = f(x)$ of one variable has two "halftangents." (See Figure 14.4.)

This means that there always exist the limits (in the sequel $h > 0$)

$$f'_-(x_0) = \lim_{h>0, h\to 0} (f_0(x_0 - h) - f(x_0))/(-h),$$

$$f'_+(x_0) = \lim_{h>0, h\to 0} ((f(x_0 + h) - f(x_0))/h).$$

The segment $[f'_-(x_0), f'_+(x_0)]$ (which usually degenerates to a point) is called *the subdifferential* of the function f at the point x_0 and is denoted by $\partial f(x_0)$. If f is constant, then $\partial f(x) \equiv 0$, and if $f(x) = ax + b$, then

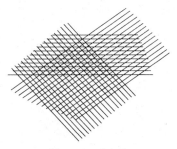

FIGURE 14.5

$\partial f(x) = a$ at each point x. In general, if $f(x)$ is differentiable at x_0, then $\partial f(x_0) = f'(x_0)$.

If a differentiable function has a local minimum at a point x_0, then, as we know, Fermat's theorem implies the equality $f'(x_0) = 0$. In this connection we noted that this is a necessary but not a sufficient condition for an extremum. If a function is convex, then its local minimum is always global, or absolute. This is one of the remarkable properties of a convex function. A necessary and sufficient condition for minimality of a convex function at a point x_0 is the relation $0 \in \partial f(x_0)$. This relation denotes a very simple condition: For a convex function to have a minimum at x_0 it is necessary and sufficient that the constant function equal to $f(x_0)$ does not lie above the graph of f.

The notion of a subdifferential can be extended to the case of a function of n variables. Unlike the derivative, a subdifferential is not a vector but a certain convex set of vectors. Also, certain formulas similar to the formulas of the differential calculus hold. One such formula is

$$\partial(f + g)(x) = \partial f(x) + \partial g(x).$$

This generalizes the formula $(f + g)'(x) = f'(x) + g'(x)$ discussed in the eleventh story. The formula for the subdifferential of a sum means that in order to find the subdifferential of the function $f + g$ at a point x, we must take the sets $A = \partial f(x)$ and $B = \partial g(x)$ and form the set $A + B$ of sums $a + b$, $a \in A$, $b \in B$.

The convex calculus consists of relations similar to the formula for the subdifferential of a sum.

The most important idea in convex analysis is that convex sets always admit dual descriptions and that for each convex set there is always a "dual" set. For example, a plane convex figure can be described as the totality of its points or as the intersection of all halfplanes (halfspaces) that contain it. (See the description of a triangle in Figure 14.5.) Similarly, every convex function can be described as the function itself or as the maximum of all affine functions that don't exceed it.

The latter description brings us to one of the most important notions of

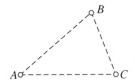

FIGURE 14.6

classical analysis, namely *the Legendre transform.*

Let $y = f(x)$ be a convex function. Its Legendre transform is the function $y = f^*(z)$, $f^*(z) = \max_x (xz - f(x))$.

Let's look at an example. Let $f_p(x) = |x|^p/p$, $p > 1$. (We recall that for $p > 1$ the function $y = f_p(x)$ is convex.) To find $\max(xz - f_p(x))$ we apply Fermat's theorem.

We have $(d/dx)(xz - f_p(x)) = z - |x|^{p-1} \operatorname{sign} x = 0$. Then $x = |z|^{p'-1}$ $\operatorname{sign} z$. Here $(p')^{-1} + p^{-1} = 1$. Also,

$$f_p^*(z) = zx - |x|^p/p = |z|^{p'} - (x|x|^{p-1} \operatorname{sign} x)/p$$
$$= |z|^{p'}(1 - 1/p) = |z|^{p'}/p' =: f_{p'}(z).$$

Thus the functions $f_p(x)$ and $f_{p'}(x)$ are dual to one another: each is the Legendre transform of the other. Also, the function $f_2(x) = x^2/2$ is "self-dual."

Here we find the reason for the Cauchy-Bunyakovskiĭ and Hölder inequalities discussed in the fifth story. Incidentally, most of the inequalities discussed in that story are related to convex analysis. For example, the arithmetic-geometric means inequality is an instance of the Jensen inequality. In fact, assume that $x_i > 0$, $i = 1, \ldots, n$. Then

$$(x_1 \ldots x_n)^{1/n} = e^{\frac{\ln x_1 + \cdots + \ln x_n}{n}} \leq \frac{1}{n}(e^{\ln x_1} + \cdots + e^{\ln x_a}) = \frac{x_1 + \cdots + x_n}{n}.$$

We've used the Jensen inequality for the convex function $y = e^x$.

Here is one more extremely important thesis of convex analysis. Consider a triangle ABC. (See Figure 14.6.) If we erase everything but its vertices, we can still reconstruct the triangle. The same can be said about a square, a rhombus, and, quite generally, about any convex polygon. They can all be reconstructed from their vertices. A vertex of a polygon can be characterized by the fact that, unlike the other points of the polygon, it is not the midpoint of a segment whose endpoints belong to the polygon.

It turns out that every bounded and closed convex set has "extremal" points, that is, points that are not midpoints of segments belonging to this set. For example, in the case of a disk, the set of extremal points coincides with its boundary circle. Also, it turns out that every convex set can be reconstructed from its extremal points.

Recall the formalization of problems of linear programming:

$$a_{01}x_1 + \cdots + a_{0n}x_n \to \min(\max),$$
$$a_{i1}x_1 + \cdots + a_{in}x_n \le b_i, \qquad i = 1, \ldots, m,$$
$$x_i \ge 0, \qquad i = 1, \ldots, m.$$

In such problems the constraints form a "polyhedron" that can be reconstructed from its set of vertices. Also, it is easy to see that the minimum (or maximum) of a linear function is attained at one of its vertices.

One of the world's best known numerical methods is the so-called *simplex method*. It enables the user to go from one vertex to another with smaller (larger) value of the minimized (maximized) function. The required extremum is found in a finite number of steps.

The theory of convex extremal problems is called *convex programming*. The following formulation describes a rather extensive class of finite-dimensional problems of convex programming:

$$(\text{p}') \qquad \begin{aligned} f_0(x) &\to \min, & f_i(x) &= 0, & i &= 1, \ldots, m', \\ f_i(x) &\le 0, & i &= m' + 1, \ldots, m, & x &\in A. \end{aligned}$$

Compared with the formulation in the beginning of this section we have the following modifications. First, this is a minimum problem; second, the function f_0 and the functions that prescribe the inequalities must be convex; third, the functions that prescribe the equalities must be affine; and, finally, there is the restriction $x \in A$. Also, the set A is supposed to be convex. In this case—for convex problems—Lagrange's words regarding his principle, words that serve as the epigraph for this story, are entirely correct. In fact, the following theorem holds:

If an admissible point $\hat{x}$ yields an absolute minimum for the convex programming problem (p') *then there are numbers* $\lambda_0, \ldots, \lambda_m$ *not all zero such that:* (A) *the nonnegativity conditions* $\lambda_0 \ge 0$, $\lambda_i \ge 0$, $i \ge m' + 1$ *hold;* (B) *the slack variable conditions* $\lambda_i f_i(\hat{x}) = 0$, $i \ge m' + 1$ *hold; and, finally,* (C) *we have the so-called minimum principle which states that $\hat{x}$ is a minimum point of the Lagrange function in the problem*

$$\mathscr{L}(x, \lambda_0, \ldots, \lambda_m) \to \min, \qquad x \in A, \qquad \mathscr{L} = \lambda_0 f_0(x) + \cdots + \lambda_m f_m(x).$$

If numbers $\lambda_1, \ldots, \lambda_m$ *are found with* $\lambda_0 = 1$ *satisfying the relations* (A)–(C) *then $\hat{x}$ is an absolute minimum for the problem.*

This theorem was proved rather recently, in 1951, by the American mathematicians Kuhn and Tucker. It plays the role of the Lagrange multiplier rule for convex programming problems.

Let's end this story by solving Dido's problem by means of the tools of convex analysis.

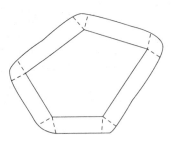

FIGURE 14.7

Let A_1 and A_2 be two plane convex figures. By $\alpha_1 A_1 + \alpha_2 A_2$ we denote the figure formed by all vectors x that are representable as $x = \alpha_1 x_1 + \alpha_2 x_2$ with $x_1 \in A_1$, $x_2 \in A_2$, and $\alpha_1 \geq 0$, $\alpha_2 \geq 0$. The sum of a polygon and a circle is shown in Figure 14.7.

Let $S(A)$ denote the area of the set A. In the middle of the nineteenth century, the German geometer Brunn proved the following important inequality:

(1) $$\sqrt{S(\alpha A_1 + (1 - \alpha)A_2)} \geq \alpha\sqrt{S(A_1)} + (1 - \alpha)\sqrt{S(A_2)};$$

then at the end of the nineteenth century H. Minkowski proved that equality in (1) is possible if and only if A_1 and A_2 are similar. Proofs of these facts and their generalizations, known as the Brunn-Minkowski inequality, can be found in [7R].

Now let A be a convex figure and B a unit circle. Then we have the following interesting formula (proved by Steiner):

(2) $$S(A + \rho B) = S(A) + p(A)\rho + \pi\rho^2.$$

In (2), $p(A)$ stands for the perimeter of A. For a convex polygon formula (2) is obvious. (See Figure 14.7.) In the general case it can be proved by passage to the limit.

We'll show how formulas (1) and (2) imply immediately the solution of the isoperimetric problem. In what follows we use the fact that $p(\alpha A) = \alpha p(A)$ and $S(\alpha A) = \alpha^2 S(A)$. We have

$$\sqrt{S(\alpha A + (1 - \alpha)B)} \overset{(2)}{=} \sqrt{S(\alpha A) + p(\alpha A)(1 - \alpha) + \pi(1 - \alpha)^2}$$
$$= \sqrt{\alpha^2 S(A) + \alpha(1 - \alpha)p(A) + \pi(1 - \alpha)^2}$$
$$\overset{(1)}{\geq} \alpha\sqrt{S(A)} + (1 - \alpha)\sqrt{S(B)} = \alpha\sqrt{S(A)} + (1 - \alpha)\sqrt{\pi}.$$

Squaring and eliminating like terms, we obtain the isoperimetric inequality $p^2(A) \geq 4\pi S(A)$ obtained in the second story. Here equality holds only if A is a circle. This is yet another solution of Dido's problem.

Now we'll go back to the eighteenth century once more.

At one point I stated that Lagrange formulated, actively applied and promoted his principle of lifting constraints in variational problems, but that he didn't prove it. The justification of this Lagrange principle for problems of the classical calculus of variations fell to mathematicians of later generations, especially those active at the end of the nineteenth century (Mayer and others). In this connection we note that Lagrange's principle follows directly from the fundamental theorems of infinite dimensional analysis.

How do we embed the classical calculus of variations in infinite-dimensional analysis? We adduce a special, but extraordinarily important, case of the general Lagrange problem.

$$I_0(y_1, \ldots, y_n, u_1, \ldots, u_r)$$
$$= \int_{x_0}^{x_1} f_0(x, y_1(x), \ldots, y_n(x), u_1(x), \ldots, u_r(x))\, dx \to \min(\max),$$

$(\mathrm{p}_1) \quad y_1' = \varphi_1(x, y_1, \ldots, y_n, u_1, \ldots, u_r),$

$$\cdots\cdots\cdots\cdots\cdots\cdots\cdots\cdots$$

$$y_n' = \varphi_n(x, y_1, \ldots, y_n, u_1, \ldots, u_r),$$

with boundary conditions $y_i(x_0) = y_{i0}$, $y_i(x_i) = y_{i1}$, $i = 1, \ldots, n$.

In this formulation the unknown functions are separated into two classes. Some $(y_1(x), \ldots, y_n(x))$ are involved in differential equations. Others $(u_1(x), \ldots, u_r(x))$ can be freely chosen. They are called *controls*. The scheme of a differential equation or a system of differential equations with control functions describes a multitude of phenomena that allow human interference.

If we assume (subject to the condition that $f_0, \varphi_1, \ldots, \varphi_n$ are continuously differentiable) that all the y's are continuously differentiable and the controls $u_1(x), \ldots, u_r(x)$ are continuous, then we can consider the mapping that associates to a set of functions $y_1(x), \ldots, y_n(x)$, $u_1(x), \ldots, u_r(x)$ the set of continuous functions

$$z_1(x) = y_1'(x) - \varphi_1(x, y_1(x), \ldots, y_n(x), u_1(x), \ldots, u_r(x)),$$

$$\cdots\cdots\cdots\cdots\cdots\cdots\cdots\cdots\cdots\cdots\cdots$$

$$z_n(x) = y_n'(x) - \varphi_n(x, y_1(x), \ldots, y_n(x), u_1(x), \ldots, u_r(x)).$$

We write this mapping in the form $z(x) = F(y_1(x), \ldots, u_r(x))$, where $z(x) = z_1(x), \ldots, z_n(x)$. The differential equation in (p_1) can be viewed as the equality $F(y_1(x), \ldots, u_r(x)) = 0$. In sum, we obtain a problem of the form

$$I_0(y_1, \ldots, u_r) \to \min(\max), \qquad F(y_1(x), \ldots, u_r(x)) \equiv 0,$$

with boundary conditions $y_i(x_0) = y_{i0}$, $y_i(x_1) = y_{i1}$, $i = 1, \ldots, n$.

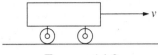

FIGURE 14.8

Lagrange's idea can be applied to this problem. Here the Lagrange function takes the form

$$\mathscr{L} = \lambda_0 I + \int_{x_0}^{x_1} (p_1(x)(y_1' - \varphi_1(x, y_1, \dots, u_r)) + \cdots$$
$$+ p_n(x)(y_n' - \varphi_n(x, y_1, \dots, u_r))) \, dx.$$

Next we must consider conceptually the problem

$$\mathscr{L} \to \min(\max), \quad y_i(x_0) = y_{i0}, \qquad y_i(x_1) = y_{i1}, \qquad i = 1, \dots, n.$$

We'll deal with this problem as if there were no constraints by writing down a system of Euler equations. In this case, the correctness of Lagrange's principle follows from the previously mentioned theorem of Lyusternik.

I'll conclude this section with an account of problems of optimal control. This extremely important chapter of the theory of extremal problems was derived from problems with technical content. Recall the simplest problem of optimal control from the first story. Suppose a cart moves rectilinearly, without friction, on horizontal rails. (See Figure 14.8.) The cart is controlled by an external force that can be changed within precribed bounds, and it must be stopped at a specified position in a given time. This simplest problem of rapid response is formalized as follows.

Let the mass of the cart be m, its initial coordinate x_0, and its initial velocity v_0. We denote the external system (the pulling force) by u, and the running coordinate of the cart by $x(t)$. Then the velocity of the cart is $v(t) = dx(t)/dt$ and its acceleration is $a(t) = d^2x(t)/dt^2$. We denote the terminal moment by T. By Newton's law, the external force $u(t)$ is equal to $m \cdot a(t)$. The constraints on the force take the form of inequalities, $u_1 \le u(t) \le u_2$. Thus we have the following formalization

$$T \to \min, \quad \frac{dx}{dt} = v, \quad \frac{dv}{dt} = u, \quad x(0) = x_0, \quad \frac{dx(0)}{dt} = v_0,$$
$$x(T) = \frac{dx(T)}{dt} = 0, \quad u_1 \le u \le u_2.$$

If there were no constraints in the form of inequalities, then the problem would fit within the classical calculus of variations. But the presence of nonstrict inequalities on the controls rules out the application of the methods of that calculus.

Problems with constraints of this type are called problems of *optimal control*. Their theory was elaborated by the Soviet scientist Pontryagin and his colleagues Boltyanskiĭ, Gamkrelidze, and Miščenko. The fundamental method of solution of such problems discovered by them is known as *the Pontryagin maximum principle*.

What sort of thing is this? A problem of optimal control is formulated almost like (p_1) except that there are additional constraints imposed on the controls. We write the latter as

$$(*) \qquad (u_1(x), \dots, u_r(x)) \in U,$$

where U is some fixed set of vectors $(u_1, \dots, u_r)$.

Let's examine a very simple example of an optimal-control problem. Let $f(x, u)$ be a continuous function of two variables. We'll consider the problem

$$(p_2) \qquad \int_{x_0}^{x_1} f(x, u(x)) \, dx \to \min, \quad u(x) \in [u_1, u_2], \quad (\Leftrightarrow u_1 \le u(x) \le u_2).$$

We'll restrict ourselves to the following special case of this problem

$$\int_{x_0}^{x_1} p(x) u(x) \, dx \to \min, \qquad -1 \le u(x) \le 1.$$

Here $p(x)$ is some continuous function on the interval $[x_0, x_1]$.

How should we proceed? It is easy to see that our integral will be least if we put $\hat{u}(x) = -1$ when $p(x) > 0$ and $\hat{u}(x) = +1$ when $p(x) < 0$, that is $\hat{u}(x) = -\operatorname{sign} p(x)$. In the general problem (p_2) we must proceed in an analogous manner, namely for each x in $[x_0, x_1]$ we must find the u in $[u_1, u_2]$ for which the function $f(x, u)$ (of u) has a minimum on this interval.

This can be formulated as the following minimum principle: For a function $\hat{u}(x)$ to be a solution of (p_2), it is necessary that

$$\min_{u \in [u_1, u_2]} f(x, u) = f(x, \hat{u}(x)).$$

Now we can explain what the maximum principle is about. Lagrange's general conception applies to the problem of optimal control except that it must be modified somewhat. If we are required to solve the optimal control problem (p_2) subject to the additional constraint $(*)$, then we must form the Lagrange function $\mathscr{L}$ (without reflecting in it the constraint $(*)$) and again consider conceptually the problem

$$\mathscr{L} \to \min, \qquad y_i(x_0) = y_{i0}, \qquad y_i(x_1) = y_{i1},$$
$$i = 1, \dots, n, \qquad (u_1, \dots, u_m) \in U.$$

With respect to y we proceed as we did earlier, that is, we form the Euler equations. With respect to u we apply the minimum principle. Since all terms involving u enter the Lagrange function with a " $-$ " sign, it is more convenient to write it as a *maximum principle*, where $(\hat{y}_1(x), \dots, \hat{y}_n(x),$

$\hat{u}_1(x), \ldots, \hat{u}_r(x))$ is a solution of the problem

$$\max_{(u_1, \ldots, u_r) \in U} (p_1(x)\varphi_1(x, \hat{y}_1(x), \ldots, \hat{y}_n(x), u_1, \ldots, u_r)$$
$$+ \cdots + p_n(x)\varphi_n(x, \hat{y}_1(x), \ldots, \hat{y}_n(x), u_1, \ldots, u_r)$$
$$= p_1(x)\varphi_1(x, \hat{y}_1(x), \ldots, \hat{y}_n(x), \hat{u}_1(x), \ldots, \hat{u}_r(x))$$
$$+ \cdots + p_n(x)\varphi_n(x, \hat{y}_1(x), \ldots, \hat{y}_n(x), \hat{u}_1(x), \ldots, \hat{u}_r(x)).$$

Now, finally, we are ready to solve Newton's problem.

$1°$ *Formalization.* Newton's problem is formalized as a problem of optimal control,

$$\int_0^a \frac{x\,dx}{1+u^2} \to \min, \qquad y' = u, \qquad u \geq 0,$$

with boundary conditions $y(0) = 0$, $y(a) = b$.

$2°$ *Necessary condition.* Application of the Lagrange principle. The Lagrange function is

$$\mathcal{L} = \int_0^a \left(\frac{\lambda_0 x}{1+u^2} + p(y' - u) \right) dx.$$

The necessary condition for y is Euler's equation

$$p' = 0 \Rightarrow p = \text{const.} = p_0.$$

The necessary condition for u is the minimality condition

$(**)$
$$\frac{\lambda_0 x}{1+u^2} - p_0 u \geq \frac{\lambda_0 x}{1+\hat{u}^2(x)} - p_0 \hat{u}(x).$$

$3°$ *Discussion.* If we suppose that $\lambda_0 = 0$, then, necessarily, $p_0 \neq 0$ (otherwise all Lagrange multipliers would be zero). If $\lambda_0 = 0$ and $p_0 \neq 0$, then $(**)$ implies that $\hat{u} \equiv 0$, that is $y(x) = \int_0^x \hat{u}(\alpha)\,d\alpha \equiv 0$. Then the required body "has no length"—it is a flat membrane. If $b > 0$, then it must be assumed that $\lambda_0 \neq 0$, and we can put $\lambda_0 = 1$. Note that the case $p_0 \geq 0$ must also be ruled out, since in that case the function $(x/(1+u^2)) - p_0 u$ is monotonically decreasing and we cannot have $(**)$.

If we study the behavior of the function $(x/(1+u^2)) - p_0 u = \varphi(u, x)$ with respect to u, then it is easy to see that for $p_0 < 0$ and small x this function attains its minimum for $u = 0$. Then the optimal control is found from the equation $-p_0 = 2ux/(1+u^2)^2$, obtained by differentiating $\varphi(u, x)$ with respect to u. The break moment ξ is determined by the fact that the function $\varphi(u, \xi)$ has two minima.

Put differently, at the break moment the following relations ($\hat{u}(\xi)$ denotes $\hat{u}(\xi + 0) \neq 0$) must hold:

$$-p_0 = \frac{2\hat{u}(\xi)\xi}{(1+\hat{u}^2(\xi))^2}, \qquad \frac{\xi}{1+\hat{u}^2(\xi)} - p_0\hat{u}(\xi) = \xi.$$

From the second equation we obtain $-\xi \hat{u}^2(\xi)/(1 + \hat{u}^2(\xi)) = p_0 \hat{u}(\xi)$, whence $p_0 = \xi \hat{u}(\xi)/(1 + \hat{u}^2(\xi))$. Substituting this relation in the first of the equations just set down we find that $\hat{u}^2(\xi) = 1 \Rightarrow \hat{u}(\xi) = 1$, for $\hat{u} \geq 0$. We then find from that same equation that $\xi = -2p_0$.

After the break the optimal control satisfies the relation

$$x = -\frac{p_0(1 + u^2)^2}{2u} = -\frac{p_0}{2}\left(\frac{1}{u} + 2u + u^3\right).$$

But

$$\frac{dy}{dx} = u \Rightarrow \frac{dy}{du} = \frac{dy}{dx}\frac{dx}{du} = u\frac{dx}{du} = -\frac{p_0}{2}\left(\frac{1}{u} + 2u + 3u^3\right).$$

Integrating this relation and bearing in mind that $\hat{u}(\xi) = 0$ for $\hat{u}(\xi) = 1$ we obtain the parametric equations of Newton's curve

$$\hat{y}(x, p) = \frac{p}{2}\left(\ln\frac{1}{u} + u^2 + \frac{3}{4}u^4\right) - \frac{7p}{8},$$

$$x = \frac{p}{2}\left(\frac{1}{u} + 2u + u^3\right), \qquad p = -p_0.$$

Instead of investigating the simplest problem of rapid response, I will refer you to the book [1R] where this problem is solved using the Lagrange principle.

It's time to end this story now. I have kept my promise and solved all problems from Part One twice. Let's take a break from formulas and have a chat.

What would I tell "the first high school student in the street" about the theory of extremal problems (recall one of the epigraphs to this story)? Surely, something along these lines: In school you learned about functions of one variable. They told you about Fermat's method of solution of extremum problems for such functions. But, in fact, there are very many problems that reduce to the minimization of functions of many variables and even functions of functions (say, curves), as in the case of the brachistochrone problem. These problems have been investigated in a chapter of mathematics called the calculus of variations. The notion of a derivative—the fundamental notion of high school analysis—was generalized in functional (infinite-dimensional) analysis, a subject that arose at the beginning of this century. Infinite-dimensional analysis makes possible a unified view of the problem of minimization of a function of one and many variables and of problems of the calculus of variations.

In this most general situation Fermat's theorem remains fully valid for problems without constraints: at an extremum, the derivative must be zero. In the case of problems of the calculus of variations, the decoded version of Fermat's theorem is a differential equation known as Euler's equation.

The number of problems without constraints is relatively small. A large part of problems with constraints can be formalized as problems with constraints in the form of equalities.

Lagrange put forward a principle for the solution of finite-dimensional problems with equalities. Its essence consists in the formation of the Lagrange function (that is, the sum of the function to be minimized and the functions that determine the equalities multiplied by undetermined coefficients) and in treating this function as if there were no constraints. (Here you could refer to the words of Lagrange in the epigraph to the twelfth story.) Lagrange's general conception remains valid for problems of the calculus of variations, as well as for problems of optimal control—a new chapter of the theory of extremal problems.

If my new student acquaintance showed further interest, I would tell him or her about the contents of Part Two of this book.

Our next story deals with some general questions.

15

More Accurately, a Discussion

All styles are fine except the boring one.

A French saying

It can hardly be denied that our elementary methods are simpler and more direct than the methods of analysis. In general, when studying some scientific problem it is better to begin with its individual peculiarities than rely on general methods.

R. Courant and M. Robbins

At the end of the century there existed a depressing tendency to turn from fundamental problems in mechanics as well as in pure analysis. Contrary to the great tradition of Jakob Bernoulli and Euler, this formalism quickly established itself in the French school and was reflected in *Analytical mechanics*.

C. Truesdell

All styles are fine except the boring one. This saying reflects the happy and life-affirming spirit of the French. Anything but boring!

In the first half of the book I tried to entertain the reader. I told fairy tales, parables, stories, and anecdotes, strained for variety, dished out romance, fable, and poetry, followed the thoughts of great men ... .

And in the second half everything changed, everything became mundane, routine, and prosaic. No stories, no poetry, no frills. Functions, derivatives, the Lagrange principle ... . Utter monotony! Formalization, necessary conditions, solution of equations—and again formalization, necessary conditions, and so on. What a bore!

Recall once more one of our epigraphs: "We must make it our goal to find a method of solution of all problems ... by means of a single, simple method." (Here d'Alembert had in mind the problems of dynamics.) But is this goal attainable? Is it not true that the truths are concealed in countless hiding places? Can we expect to discover them by means of a single key or even a small bunch of keys?

Courant and Robbins disagree with d'Alembert; "When studying some scientific problem it is better to begin with its individual peculiarities than rely on general methods," they say. The well-known contemporary mechanician Truesdell contrasts the formalism of Lagrange with the great tradition of Euler and openly sides with Euler (see the epigraph to this story).

Who is right—Euler or Lagrange, d'Alembert or Truesdell?

We will ignore the apparent obviousness of the answer and dare to ask: Is poetic romance better than boring monotony?

You can't brush these questions aside easily, especially if you learn or teach. How can you learn and what should you teach? The concrete or the abstract? Problems or general principles?

These are the questions I propose to discuss in this "story."

Some of the questions posed here may seem trivial. I'll begin with the simplest one—the one about romance and routine.

Romance captivates us. We are attracted by mountains, icy deserts, stormy waves, danger, and risk. We revere great heroes—travellers who discover new lands, mountain climbers who scale inaccessible peaks, and brave and daring seafarers.

There was a time when reaching the North or South Pole was an enterprise for heroes. Once I happened to talk to a man who reached both poles. As a young man he liked to travel. Then he suffered from an eye disease and could no longer carry heavy backpacks or make long marches on foot. It was after he contracted the disease that he stayed at the poles. To do this he didn't have to freeze, to get over icefields, or fall through polynias.* He flew there in an airplane.

To this very day you can try to reach one of the poles by yourself or with a group of friends, using dog teams, or skis, or in a balloon, surmounting difficulties, with romance and risk.

But there is also the other way of getting there—by plane. When did easier travel begin? Was it not when the wheel was invented? And then came the

---

*Open water in an icefield, which is usually frozen over. (Tr.)

cart, the steam engine, the railroad, the automobile, and, finally, the plane. The plane allows anyone to reach the poles without poetry and romance.

Man cannot advance without heroism, without obsessive preoccupation, without risk, without romance and poetry. But then, unavoidably, the time comes when the alluring and distant goal becomes accessible to all, when it has been truly mastered. Then no one does anything heroic to reach the goal. Someone prepares the plane for a flight and does ordinary, everyday work. Then comes the takeoff order. At the airport there are no escorts and no orchestras. There are just people doing their work. The plane takes off and then lands at the pole. But it is only when such a flight becomes ordinary, routine, or commonplace, that we can say that the pole has been mastered.

Can anyone doubt that at some future time, when mankind will have mastered its burden of aggravating problems, it will make sure that everyone can stand on Mount Everest without clambering up its face and gasping for breath?

Progress in life and in science combines the efforts of pioneers and the steady forward movement of our whole civilization. In time, this movement makes accessible to all people the goals that were attained earlier only by heroes through suffering and sacrifices.

In science, as in life, you can pursue two different aims. You can train to try to scale an inaccessible peak by yourself, without special equipment. But you must also take part in the collective effort that secures the steady movement of the civilization by building roads and communication lines to the peaks. And therefore you should learn both approaches.

This brings us to the topic of what and how to teach. There is learning for development (at preschool age), for general education (in school), and for professional education (at colleges and universities). Each of these stages must be considered separately. One could well ask: Has this not yet been thought through? After all, this is one of the most fundamental questions, of relevance to each individual and to all mankind.

There was a time when I thought that all questions had been answered and that all has been known for a long time—on earth, in heaven, in science, and in life. After all, there has been life on earth for so long, and there have been so may wise men!

In this book I have tried to present a chapter of mathematics from its origin to its present state. Let's glance once more at the past. Did this chapter begin a long time ago? Yes and no. Twenty five centuries is, of course, a long time. On the other hand, think of a selection of people, one per generation, five per century. This gives 125 people from the present generation to Aristotle. How small a number!

We are still very young. Mankind has just begun to master the world. Strictly speaking, sciences in the modern sense of the word began some 300 to 400 years ago. We already know so much, yet so little! Your grandparents may have known people who were born in the age of the cart. Virtually before our eyes, in little more than a hundred years, man has mastered all that now fills our life—steamships, railroads, the telegraph, the telephone, the automobile, the plane, the TV set, artificial satellites. Mankind has not yet found the time to think through in detail some of the most important questions in life. In particular, there is no clear answer to the question of what to teach. But regardless of how the question of the content of education will be answered in the future (should one, in addition to the usual subjects, teach very early the handling of computers and word processors, car driving, shorthand and typing, editing, and so on—or should one not?), I have no doubt that one should teach mathematics for at least two reasons—to train the mind and to make possible the understanding of the structure of the world.

The debate over how to teach mathematics has been alive throughout our century. At the beginning of this century the best mathematicians—such as Klein, Borel, and Hadamard—took part in this debate. In order to delineate one of the issues of this unceasing debate I will quote at some length from Dieudonné, one of the most distinguished French mathematicians of our time:

> Please look objectively at the following topics that take up
> most of the time in school mathematics:
>
> I. "Ruler and compass" constructions.
> II. Properties of "traditional" figures, such as triangles, quadrilaterals, circles, and systems of circles—with all the refinements accumulated by generations of "geometers" in search of suitable examination questions.
> III. A whole psalter of "trigonometric formulas" and their kaleidoscopic transformations that make possible the finding of splendid "solutions of problems" on triangles and—please keep this in mind—"in a form convenient for taking logarithms ... "

Dieudonné goes on to say that no one encounters anything like this in life. He runs down "old school mathematics" with the same vigor and insistence:

> The question arises of whether it is more important for the
> builder to know that the altitudes of a triangle intersect in
> one point or to know the principles of the theory of strength
> of materials?

Dieudonné thinks that one should teach principles, and only principles. (Read the introduction to Dieudonné's book *Linear algebra and elementary geometry*. There you will find many other interesting things.)

I learned "the old way" and I can add a great deal more to Dieudonné's list, for example, arithmetical problems of the "pool-filling" type, endless arithmetical examples involving addition of ordinary and decimal fractions, and so on.

You might again think that there is nothing more to discuss, that Dieudonné is obviously correct. It is indeed the case that

> Trigonometric formulas are indispensable for representatives of three thoroughly respectable professions: 1. for astronomers; 2. for surveyors; 3. for writers of trigonometry textbooks,

and for no one else. Why then befuddle the poor schoolboy?

And yet there is something that will not allow me to concede the correctness of these words. "Education" is a very complicated notion. It involves not only the acquisition of knowledge but also training in how to think. It appears that for two centuries (or more) pool-filling problems, construction problems, problems on triangles, and transformations of trigonometric formulas served a vital purpose—they provided food for the mind, they taught exactness and accuracy, they taught reasoning, the search for truth, the surmounting of difficulties, the trying out of different roads leading to some objective, and the reaching of that objective. They imparted the joy of achievement and a sense of beauty. In brief, they modeled creativity. With what do we replace all this? And is it worth it?

It is absolutely necessary to retain these elements of creativity. The materials may, conceivably, be changed but these elements must be retained. *The only way to teach thinking is with concrete "special" problems* and not with general principles alone. It seems to me that one must provide the opportunity for solving Heron's problem Heron's way, for looking at Euclid's problem with Euclid's eyes, for experiencing the difficulty of Archimedes' problem after reaching the level of his technical means, for trying to solve Steiner's problem by oneself.

I wrote Part One because I think that extremum problems provide wonderful material for teaching thinking, inventiveness, scientific flexibility, and the overcoming of intellectual difficulties.

But I have no doubt that one must also teach the understanding of the essence of things, general principles and laws, both in the natural sciences and in life. Thus it seems to me that everyone must be familiar, at least in some general way, with the basics of mathematical analysis, because mathematical analysis is an inseparable component of the natural sciences.

There is something else that is important. It is important to realize and to understand the unity and the variety of the world. Electricity, light, heat, fluid motion, the motions of the planets—these are all different, but they all have features in common. Nature is "controlled" by general laws, all is connected, and all gravitates to oneness. And there is yet another idea that I think is important, and that is to realize that there exist "general principles." Such principles exist in mathematics as well. Lagrange, d'Alembert, and many other great scientists tried to attain an understanding of that "very essence" that unifies the uncoordinated phenomena in the world. One such principle—Lagrange's principle—was discussed in Part Two of this book.

Recall that we tried to solve each problem twice, once in Part One "beginning with its individual peculiarities," and a second time in Part Two "by relying on general methods" (I am quoting once more Courant and Robbins). Courant and Robbins arrived at the conclusion that "elementary methods are simpler and more direct than the methods of analysis." Does this hold?

It seems to me that in this dispute the general Fermat-Lagrange method is certainly second to none. True, it is sometimes defeated—for example, in the problem of the triangle of least perimeter (of course, it is possible that my solution is not optimal). It is also possible that some of my readers will declare other brilliant elementary solutions the victors. But it can hardly be denied that this method fought nobly, never once refused the challenge of single combat, and always brought us to our goal. And in some cases its victory was undeniable.

We can see in all this a definite pattern that often accompanies man's quests. A glimmer of truth appears in the dark, man wanders a long time, makes his way through an impenetrable thicket, and then it turns out that all the long and painful searches have been in vain and the road to the goal is a short one. But have the efforts really been in vain? Think about it. I wanted my book to give you a chance to retrace the thorny road of the search for truth.

There is one more aspect of the Courant and Robbins position that is debatable. At the time when Courant and Robbins wrote their book, much was made of the contrast between elementary and higher mathematics. This contrast is 300 years old. The boundaries of elementary mathematics are determined by the proximity of mathematical analysis. All that was created before the birth of mathematical analysis, and a great deal of what turned up since then but doesn't use its methods and constructions, is classified as elementary mathematics. Mathematical analysis itself is sometimes called "higher mathematics." At one time it was thought that higher mathematics contains something "supernatural," something beyond the understanding of ordinary people, something truly "higher," something that cannot be discussed in school. This is not true at all.

Mathematical analysis is a perfectly natural, simple, and elementary discipline, not one iota more abstruse, complex, or "higher" than, say, "elementary" geometry. Nowadays, insistence on the contrast between elementary mathematics and mathematical analysis is counterproductive. There is no need to display tremendous cleverness out of fear of using the properties of the derivative.

Introduction of the elements of mathematical analysis into school programs is bound to lead to a restructuring of other areas of mathematical education as well. The content of competition problems, of the work of mathematical circles, and of mathematical olympiads is bound to change. By now it is impossible to ignore the fact that a high school student must know something of the higher mathematics to which no access was available earlier. In the thirteenth story, when we solved anew the problems from Part One and discussed certain questions that could not be accommodated on the elementary level, I tried to demonstrate the possibilities hidden in the simplest tools of mathematical analysis.

In this connection you should bear in mind that as soon as you have mastered the very basics of mathematical analysis, you can try to approach many contemporary problems. In the fourteenth story I tried to illuminate the road traversed in the theory of extremal problems from the time of Newton, Leibniz, Euler, and Lagrange to the present day. I wanted to show that, in reality, the distance from Newton to ourselves is not far.

I took as the epigraph for Part One of this book the words of Bertrand Russell, in which he contrasted mathematics with art. Before the beginning of the fourth story I quoted the words of G. H. Hardy in which he puts the Scientist above the Poet. I won't get into this argument. Science and art unite in the combined notion of the worth of man's reason. I hope my readers will apprehend a small fragment of the history of mathematical science as part of our general cultural heritage.

Bibliography

1. W. Blaschke, *Griechische und anschauliche Geometrie*, München, 1953.
2. ──, *Kreis und Kugel*, Leipzig, 1916, Berlin, 1956.
3. R. Courant and H. Robbins, *What is mathematics?*, Oxford Univ. Press, Oxford, 1978.
4. H. S. M. Coxeter, *Introduction to geometry*, Wiley, New York, 1961.
5. *The geometry of René Descartes*, Dover, New York, 1954.
6. S. G. Gindikin, *Tales of physicists and mathematicians*, Birkhäuser, Boston, 1988.
7. G. H. Hardy, J. E. Littlewood, and G. Pólya, *Inequalities*, Cambridge Univ. Press, Cambridge, 1952.
8. J. Kepler, *Nova stereometria doliorum vinariorum*. In: Johannes Kepler Gesammelte Werke, Munich, 1937.
9. A. Koestler, *The sleepwalkers*, Macmillan, New York, 1968.
10. *The mathematical papers of Isaac Newton*, edited by D. T. Whiteside, Cambridge Univ. Press, Cambridge, 1967.
11. I. Niven, *Maxima and minima without calculus*, Dolciani Mathematical Expositions No. 6, 1981.
12. H. Rademacher and O. Toeplitz, *The enjoyment of mathematics*, Princeton Univ. Press, Princeton, 1966.
13. I. M. Yaglom and V. G. Boltyanskiĭ, *Convex figures*. Holt, New York, 1961.
14. H. Zeuthen, *Geschichte der Mathematik im Altertum und Mittelalter*, Copenhagen, 1896.
1R. В. М. Алексеев, В. М. Тихомиров, С. В. Фомин, Оптимальное управление.—М.: Наука, 1979.
2R. В. М. Алексеев , Э. М. Галеев, В. М. Тихомиров, Сборник задач по оптимизации.—М.: Наука, 1984.
3R. Ю. А. Белый, Иоганн Кеплер.—М.: Наука, 1971.
4R. В. Г. Болтянский, И. М. Яглом, Геометрические задачи на максимум и минимум.—В кн.: Энциклопедия математики, кн. V.—М: Наука, 1966, с. 270–348.
5R. С. И. Зетель, Задачи на максимум и минимум.—М.—Л.: Гостехиздат, 1948.
6R. Д. А. Крыжановский, Изопериметры.—М.—Л.: ОНТИ, 1938.
7R. Л. А. Люстерник, Выпуклые фигуры и многогранники.—М.: Гостехиздат, 1956.
8R. Е. А. Предтеченский, Кеплер. Его научная жизнь и деятель Ность.—Петроград: Изд-во Гржебина, 1921.
9R. Л. В. Тарасов, А. Н. Тарасова, Беседы о преломлении света.—М.: Наука, 1982.—Библиотечка. «Квант», Вып. 18.
10R. И. Ф. Шарыгин, Задачи по геометрии. Планиметрия, изд 2-е.—М.: Наука, 1986.—Библиотечка «Квант», Вып. 17.
11R. И. Ф. Шарыгин, Задачи по геометрии, изд 2-е.—М.: Наука, 1984.—Библиотечка «Квант», Вып. 31.
12R. Д. О. Шклярский, Н. Н. Ченцов, И. М. Яглом, Геометрические неравенства и задачи на максимум и минимум.—М.: Наука, 1970.